unit 14

Pattern Specification and Morphogenesis

Contents

Table A Scientific terms and principles used in Unit 14 4

Study guide for Unit 14 5

1 **Introduction** 6

2 **Pattern specification** 6
2.1 How do you make a pattern? 7
2.2 A one-dimensional example of pattern specification 8
2.3 A model of pattern specification 9
2.3.1 The regeneration of a perturbed pattern 9
2.4 From theory to experiment 10
2.5 The segmental gradient and positional information 14
Summary of Section 2 15
Objectives and SAQs for Section 2 15

3 ***Drosophila* and development** 17
3.1 The biology of *Drosophila* 18
3.2 Imaginal disc transplantation 19
3.2.1 Competence of discs 19
3.2.2 Regeneration and duplication of disc halves 20
3.3 Gradients in imaginal discs 22
3.4 Models and development 22
Summary of Section 3 23
Objectives and SAQs for Section 3 23

4 **Cells and morphogenesis** 24
Summary of Section 4 26
Objectives and SAQs for Section 4 27

5 **Morphogenesis in cell populations** 27
5.1 Cellular reaggregation 27
5.2 The cellular basis of sorting out 28
Summary of Section 5 28
Objective and SAQs for Section 5 29

6 **From pattern specification and cell properties to morphogenesis** 29

7 **Gastrulation** 30
7.1 Sea-urchin gastrulation 30
7.2 Amphibian gastrulation 32
Summary of Section 7 35
Objectives and SAQs for Section 7 36

8 **Embryonic induction** 37
8.1 The chemical identity of the inductor 38
8.2 A further complicating factor: a whole range of inductors 39
8.3 An *in vitro* model for induction 40
8.4 Will the real inductor please stand up? 42
Summary of Section 8 42
Objectives and SAQs for Section 8 43

(*cont.*)

9 Genes and the control of development 44
9.1 Mutants and development 44
9.2 Mutations occur in cells 46
9.3 Cells and their lineages in development 47
9.3.1 Clonal analysis 48
9.3.2 The uses of clonal analysis 49
9.3.3 Compartmentation in development 49
9.3.4 Chimaeric mice 51
9.4 The genetic control of pattern formation 51
9.4.1 The *engrailed* gene 52
9.4.2 Homeotic genes 52
Summary of Section 9 54
Objectives and SAQs for Section 9 54

10 Summary and conclusions to Unit 14 56

Objectives for Unit 14 57

ITQ answers and comments 58

SAQ answers and comments 58

References and further reading 59

Acknowledgements 59

Table A Scientific terms and principles used in Unit 14

Assumed knowledge†	Introduced in an earlier Unit	Unit	Introduced or developed in this Unit	Page
abdomen	allele	11	amoebocyte	24
adrenal gland	animal pole	11	ancestry of cells*	47
agar	blastomere	11	archenteron*	31
adult	blastula	11	*bithorax* genes	53
compound eye	blastocoel	1, 11	blastema	21
diffusion	Britten and Davidson model	12 & 13	blastoderm stage*	18
Drosophila	cleavage	11	blastopore*	34
epidermis	cuticle of insects	2	bottle cells*	34
epithelium	determination	11	cell adhesiveness*	28
femur	ectoderm	1, 11	cell alignment	25
gene[2]	endoderm	1, 11	cell autonomous mutant*	52
histology	embryonic induction	11	chemotactic theory	28
kidney	gastrulation	1, 11	chimaera*	51
liver	gastrula	1, 11	clone*	47
larva	haltere	2	clonal analysis*	48
lineage	*Hydra*	SLO	collagenase	41
model	haemocoel	1	compartments*	49
mutant[1]	instar	2	compartment boundaries	49
muscle	Jacob and Monod theory	12 & 13	competence*	19
mammary tissue	membrane	5	conditioned medium	38
nerves	metamorphosis	2	contact guidance*	25
pharynx	metazoan	1	contact inhibition*	26
physiology	morphogenesis	11	cortical cytoplasm*	18
pancreas	microfilaments	5	de-differentiation	21
salivary gland	mesoderm	1, 11	dorsal lip of the blastopore*	38
solvent	notochord	3	*Drosophila*	17
thorax	pseudopodia	1	epithelium–mesoderm interactions*	40
thymus	protein	5	ectoderm–mesoderm interactions*	40
tibia	protistan	1	embryonic induction*	37
	polysaccharide	5	*engrailed* gene*	52
	regulation	11	expression of genes	47
	ribonucleoprotein	5	fate maps*	20
	vegetal pole	11		

* These terms must be thoroughly understood—see Objective 1.

† Most of these terms are explained in the Science Foundation Course. Those that are of particular importance to the understanding of this text appear with a superscript number and have a full reference at the end of the Unit.

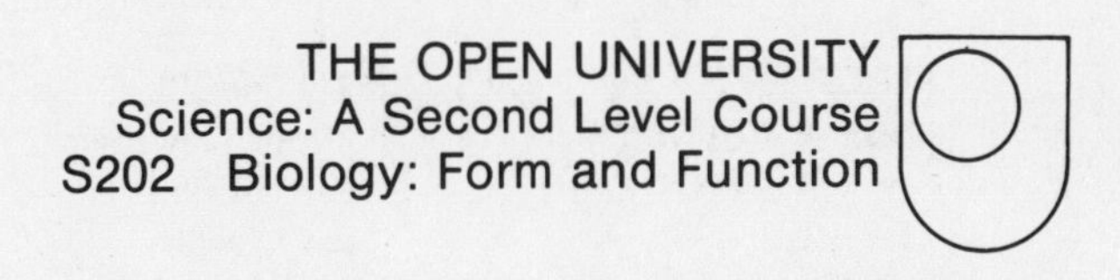

Development II

unit 14
Pattern Specification and Morphogenesis

unit 15
Chicken or Egg?

Prepared by the S202 Course Team

The S202 Course Team

This Course has been prepared by the following team:

Peggy Varley (*Chairman and General Editor*)

Hendrik Ball (*BBC*)
Gerry Bearman (*Editor*)
Mary Bell
Eric Bowers
Bob Burgoyne
Ian Calvert
Bob Cordell
Norman Cohen
Peter Cole (*BBC*)
Baz East (*Illustrator*)
Vic Finlayson
Anna Furth
Denis Gartside (*BBC*)
Lindsay Haddon
Robin Harding
Stephen Hurry
David Kerrison (*BBC*)
Aileen Llewellyn (*BBC*)
Pam Mullins
Pat Murphy
Seán Murphy
Pam Owen (*Illustrator*)
Phil Parker (*Course Coordinator*)
Ros Porter (*Designer*)
Rob Ransom
Irene Ridge
Steven Rose
Ian Rosenbloom (*BBC*)
Jacqueline Stewart (*Course Editor*)
Mike Stewart
Jeff Thomas
David Tillotson (*Editor*)
Charles Turner
Sue Turner

The Open University Press,
Walton Hall, Milton Keynes.

First published 1981

Designed by the Graphic Design Group of the Open University.

Typeset by Santype International Limited, Salisbury, Wilts, and printed by W. & J. Linney Ltd, Mansfield, Notts.

ISBN 0 355 16035 2

This text forms part of an Open University course. The complete list of Units in the Course is printed at the end of this text.

For general availability of supporting material referred to in this text please write to: Open University Educational Enterprises Limited, 12 Cofferidge Close, Stony Stratford, Milton Keynes, MK11 1BY, Great Britain.

Further information on Open University courses may be obtained from the Admissions Office, The Open University, P.O. Box 48, Walton Hall, Milton Keynes, MK7 6AB.

1.1

Introduced or developed in this Unit	Page	Introduced or developed in this Unit	Page
fibroblast	24	neurula*	37
field*	7	normal and abnormal inductors*	39
French flag model*	9	notochord*	38
gastrulation*	30	ommatidium (*pl.* ommatidia)	48
glycosaminoglycans	41	organizer*	38
gradient*	7	pattern	6
Gustafson & Wolpert's theory	31	pattern specification*	6
heterocyst	8	pleiotropy	44
homeotic mutant*	45	polarity*	11
hyaline layer	31	positional information*	14
hyaluronidase	41	presumptive	14
imaginal disc*	18	primary invagination*	32
inductor*	38	pro-heterocyst	8
inherently precise machine	26	regeneration by growth*	10
inhibitor gradient	8	regeneration by remodelling*	9
insect cuticular patterns	10	regulative development*	7
leading lamella*	24	Saxon and Toivonen's interpretation of induction	39
mesodermalizing agent*	39	secondary inductions	40
Millipore filter	40	secondary invagination*	32
Minute mutant	49	serial culturing	21
mirror image duplicate*	20	sorting out*	27
mosaic development*	7	surface coat	32
mutants	44	*talpid* mutants	46
morphogen*	9	threshold inhibition	8
morphogenesis	6	timing hypothesis	28
neural crest*	40	transdetermination*	21
neuralizing agent*	39	unconditioned medium	38
neural plate*	35		

Study guide for Unit 14

Morphogenesis is the process by which shapes and patterns are formed during development. Pattern specification is the mechanism by which cells acquire their positional cues in a developing system. Because these two topics are so closely related, they cannot be treated separately. Hence this is a long Unit, representing almost two weeks' work. We begin by looking at what is needed to specify a pattern, and then consider the properties of individual cells that are important in morphogenesis. Next, the more complex aspects of multicellular morphogenesis are examined, and the final section discusses the genetic control of pattern specification and morphogenesis.

With some exceptions, we concentrate here on animal development. Pattern specification and morphogenesis in plants are discussed in detail in Units 29–31. The reason for this separation is that although there are factors common to both plant and animal development, there are also large differences: unlike the animal cell, the plant cell is enclosed in a rigid wall and is non-motile, and animals pass through a distinct embryonic phase, whereas plants have embryonic structures—the apical meristem of shoot and root—which are capable of further growth and development throughout the life of the organism.

The techniques of molecular biology have given valuable insight into the mechanism of cell differentiation as you have seen in the previous Unit. However, knowledge of how differentiation is spatially arranged to give the changes of shape that make up morphogenesis is still very limited. You may feel rather overwhelmed by the number of experimental systems and theories described here, and you should bear the objectives for the Unit in mind as you read through the text. Plan your approach to the Unit by carefully reading the study comments below, and those given at the beginning of each main Section.

Start by looking through the Unit and studying the contents list. You should then be able to make a preliminary assessment of the amount of work involved and plan your study time accordingly. Pay particular attention to the general principles outlined in each Section: your attention is drawn to these at the beginning of certain Sections.

There are two television programmes associated with the Unit, with the general title *Patterns in Development*. The first deals with the role of gradients in development, discussed in Sections 2 and 3. The second is concerned with cell movement in development (Sections 4–7). You should attempt to read these Sections before watching the television programmes. There is also an audiocassette band on comparative embryology, which should preferably be listened to after completing the Unit.

Occasional reference is made in this text to *A Survey of Living Organisms**, and to *The S202 Picture Book*†. Have these to hand as you work through the Unit.

Unit 14 is a long Unit, which will take more than one week's study time. The last Unit of the developmental biology Block, Unit 15, is very short. We suggest that you therefore study Units 14 and 15 over a two week period but spend the majority of the time on Unit 14.

1 Introduction

In the preceding two Units, we looked at cell differentiation and how it may be controlled. We considered how the genetic information is transcribed and translated at particular times during development. Besides this temporal control system, organisms also change their shapes and generate new *patterns* as they develop. A pattern is a spatially non-homogeneous region in a developing organism, and the process of formation of new patterns and shapes during development is *morphogenesis*. It is not difficult to visualize the sort of biochemical mechanisms that might control differential gene transcription, although we cannot as yet describe differentiation in precise molecular biological terms. How cells in a developing organism communicate spatially one with another to produce new structures is much more difficult to comprehend, however. This Unit should give you some insight into the arguments and experiments involved in this controversial field.

pattern

morphogenesis

The Unit is laid out in the following manner. The first questions to be answered are: How do cells know where they are? Is there a 'map' for developing cells to follow? This is the problem of *pattern specification*. Cells must interpret this map according to their make-up, and so the next topic for study is the sorts of cell properties that may be important in morphogenesis. This leads on to the interaction of cells one with another in morphogenesis, and we look at several examples of cell cooperation during development. Finally, we turn to the control of these various events and discuss genes and morphogenesis.

pattern specification*

2 Pattern specification

Figure 1 shows some examples of biological patterns. Patterns may be overt, that is, obviously differentiated (like the stripes on a zebra's pelt) or they may be covert (where the individual cells of a region may be differently determined, but show no easily measurable differences at the time during development when it is observed). The 'developmental region' could be anything from the whole embryo (the distribution of the individual organs would be the pattern), to a small group of cells destined to form an organ or a tissue making up part of an organ. A pattern could even be within a single cell, for example, the protistan *Tintinnopsis* (Unit 1, Section 3.3.1) or a bacterium. We limit our study to multicellular animals.

* The Open University (1981) S202 SLO *A Survey of Living Organisms*, The Open University Press. This text forms part of the supplementary material for the Course.

† The Open University (1981) S202 **PB** *The S202 Picture Book*, The Open University Press. This contains colour and half-tone illustrations for the Course.

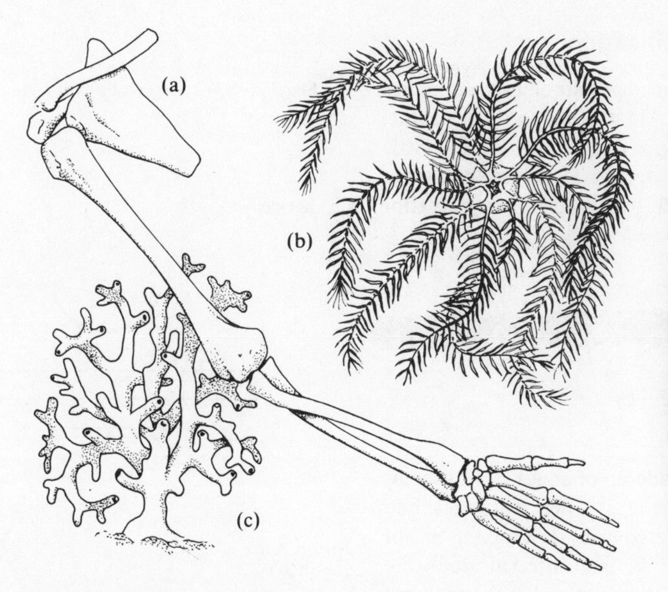

FIGURE 1 Patterns in biological systems. (a) Bones of human arm; (b) feather star; (c) branching sponge; (d) compound eye of insect.

2.1 How do you make a pattern?

The building blocks of metazoan animals are cells, and it is important to keep a mental picture of individual cells cooperating spatially to produce structures during development. How can ordered diversity be generated from a mass of identical cells to produce a particular pattern? Theoretically all have the same genetic information. In the preceding Unit, we looked at the *temporal* changes that go on in cells, that is, how particular genes may be switched on and off. Here we are concerned with *spatial* changes, and there are two ways in which patterns arise. The first involves little or no communication between neighbouring cells and is called *mosaic development*. Here cells are like pieces of a mosaic: remove one and a specific gap is left. Alternatively, development may be *regulative*, and here removal of cells is regulated by the remaining cells, so that new cells are formed to compensate for the missing portion. Mosaic development is seen, for example, in the early growth of molluscs and flatworms; regulative development is found, for example, in sea-urchins and some insects.

mosaic development*
regulative development*

Although whole groups of animals often show similar patterns of regulation or mosaicism, there are no hard and fast rules. Insect embryos are often mosaic, while other invertebrates regulate. Amphibians show striking regulation in the regeneration of limbs, but mammals do not.

□ What essential properties must cells of a regulating organism have?

■ They must be able to change their differentiated state to allow the missing cell types to be formed. They must also be capable of dividing.

An area of embryonic tissue that takes part in a particular morphogenetic event is usually called a '*field*', and the first necessity for regulative systems is to have some form of communication between the cells of the field. If the cells do not 'know' what differentiated state their neighbours are in, it will be difficult to draw boundaries between regions. Some signalling mechanisms must be imposed on the system, and cells could respond to this signal in some way.

field*

If the system is a purely mosaic one, then an alternative explanation to signalling is, theoretically, possible. Cells might arise from a basic group of cells already arranged in a particular pattern, and as each new cell is laid down, it 'fits' against its neighbours according to instructions inherent in the cell itself. Such a mechanism has been proposed to explain early cleavage in the egg of the snail *Limnaea*.

This kind of explanation apart, the simplest way of setting up a pattern is to invoke the presence of a *gradient* of some kind. The study of pattern specification is pervaded by the idea of gradients, and it is important that you understand their basic features. The first TV programme associated with this Unit will deal with gradients in specific developmental situations. Although gradients could be formed from any quantifiable factor that can change its level over distance (e.g. pressure or electrical impulse), most models have been framed in simple biochemical terms, suggesting that cells in the field respond to particular levels of some diffusible chemical of low relative molecular mass. How might this kind of pattern specification work in practice?

gradient*

2.2 A one-dimensional example of pattern specification

Anabaena is a blue-green alga (Unit 1, Section 2) which consists of an unbranched filament of cylindrical cells, as shown in Figure 2. Most of the cells are rather squarish in shape when seen under the microscope, but at regular intervals, another cell type—rather larger and more rounded—occurs. These are specialized cells called *heterocysts*, which are concerned in some way with nitrogen fixation (Unit 1, Section 2). Heterocysts do not divide.

See *The S202 Picture Book* for *Anabaena*.

heterocyst

FIGURE 2 A single filament of *Anabaena* (×200).

Normally, the interval between two heterocysts is made up of about eight vegetative cells, but as the cells divide, the interval widens until when it has reached about 16 cells, a new presumptive heterocyst (*pro-heterocyst*) develops in about the middle of the interval. This process is repeated in each interval along the filament, and in this way, a spaced heterocyst pattern is maintained as the filament grows. Obviously some mechanism operates to ensure that new heterocysts do not develop near existing heterocysts. It appears that there is an inhibitory zone around each heterocyst; further heterocysts differentiate only when they lie outside these zones.

Could this simple pattern formation mechanism be explained by a gradient? Wilcox and Mitchison proposed that heterocysts might produce an inhibitory substance that travels outward from each heterocyst along the filament. As it travels, it may be absorbed or destroyed by the surrounding vegetative cells. There would thus be a *gradient of inhibitor concentration* leading away from each heterocyst. The concentration of the proposed inhibitor substance would be lowest midway between each pair of heterocysts. There may be a *threshold inhibitor* concentration; any cell experiencing an inhibitor concentration below this will 'switch on' and differentiate into a heterocyst. Of course, at this stage, this is just a model that is used to interpret the observations.

Suppose we take a chain of about eight cells between two heterocysts (see Figure 3a). The curve represents an inhibitor concentration in the cells in this filament. The heterocysts are very close together, so that all the cells in the interval between them have a high, above-threshold inhibitor concentration. As these cells divide, the interval widens and the concentration of inhibitor in the middle of the interval drops until eventually it falls below the threshold in some of these cells (see Figure 3b). This is the signal for those cells to begin differentiation into a heterocyst (Figure 3c) and to start producing inhibitor (Figure 3d).

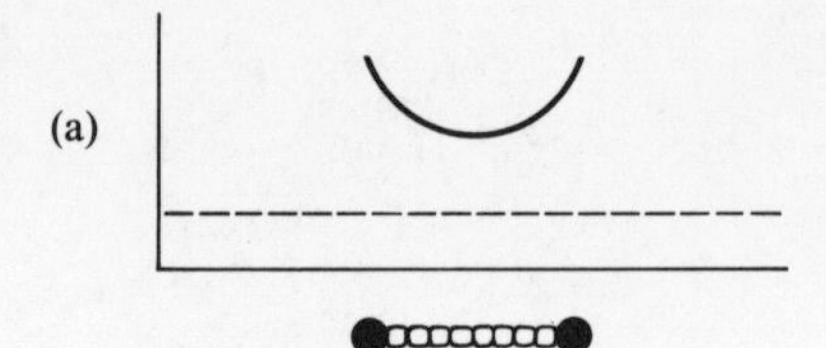

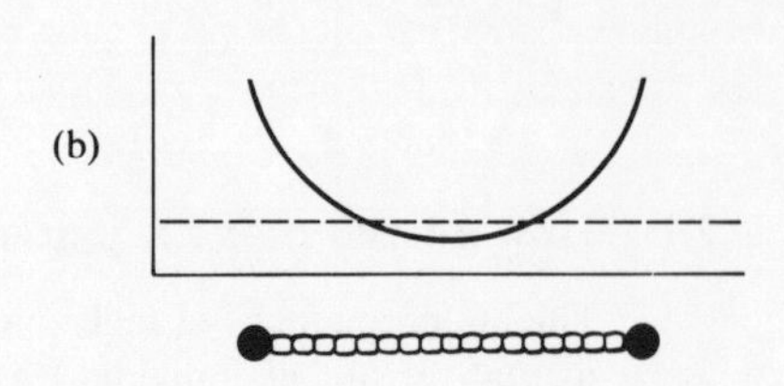

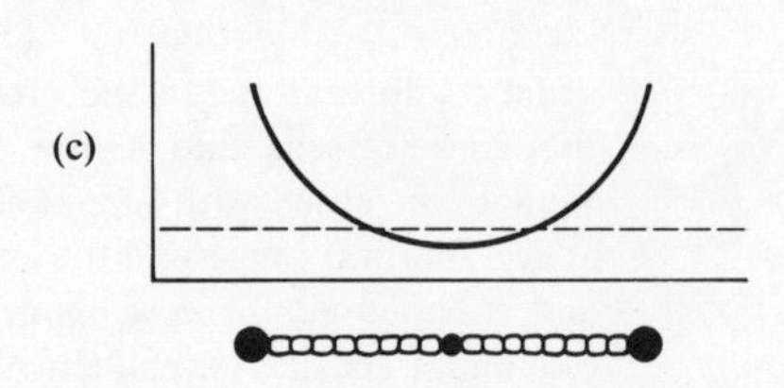

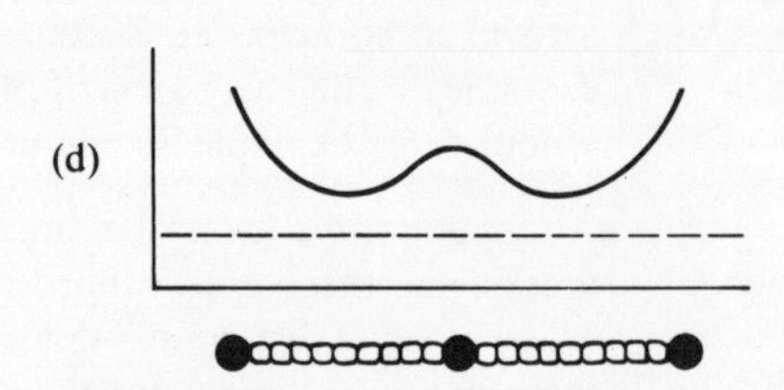

FIGURE 3 New heterocyst formation in *Anabaena* caused by fall in inhibitor concentration. (a) Postulated curve (gradient) of inhibitor in filament of *Anabaena* cells. (b) Schematic representation of change in level of postulated inhibitor during growth of *Anabaena* filament. (c) and (d) Schematic representation of a heterocyst forming at sub-threshold levels (c) and inhibitor formation by new heterocyst (d). The broken horizontal line represents the threshold concentration of the inhibitor.

A problem arises with such a simple mechanism, for it is difficult to see how it could be made sufficiently accurate to ensure that single cells. rather than groups of cells, are picked to differentiate and thus to prevent a whole group of adjacent heterocysts from being formed. The organism has two methods of overcoming this problem. The first of these concerns cell divisions. Wilcox and Mitchison observed that all cells divide *asymmetrically* (see Figure 4). Each cell gives rise to two daughter cells—a long one and a short one. The importance of this for the pattern is that heterocysts develop only from the smaller daughter cells of these asymmetrical divisions. This asymmetry results in a single heterocyst being picked in most cases. There is, however, a second, competitive mechanism that operates to make sure that if two cells do start to differentiate when they are too close together, only one of them eventually develops. It seems that during the early stages of development, a pro-heterocyst can be inhibited (or 'switched off') if the inhibitor level in it becomes too high. If two pro-heterocysts are developing too close together, each is liable to be inhibited by inhibitor produced by the other—a sort of competition occurs for the one available 'slot'. Experimental results demonstrate this rather dramatically. A pro-heterocyst will inhibit itself if it is isolated from the filament together with a small number of vegetative cells (see Figure 5). The pro-heterocyst, which would normally have developed into a heterocyst within a few hours, divides and reverts to the vegetative state. This is because these few vegetative cells cannot absorb all the inhibitor produced by the pro-heterocyst; consequently its concentration builds up until it inhibits the pro-

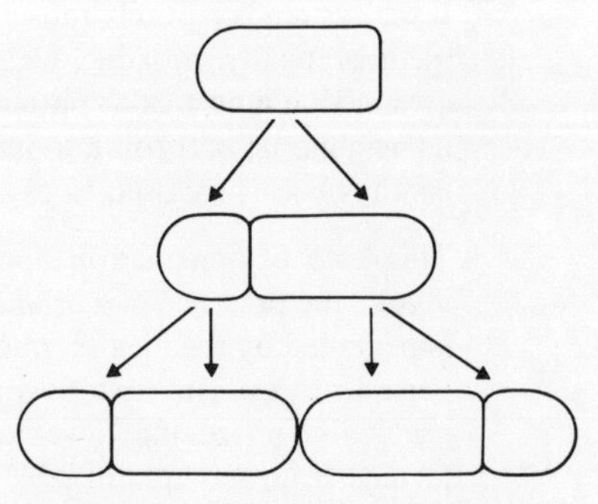

FIGURE 4 Diagram of asymmetrical cell division.

heterocyst itself. There must be a clearly defined inhibitory zone, which may take the form of a chemical gradient, around heterocysts. Thus, asymmetrical division and competition seem to be the minimum requirements for the setting up and maintenance of the developmental pattern in *Anabaena*.

Pattern formation in *Anabaena* is also considered in the TV programme *Patterns in Development: Gradients.*

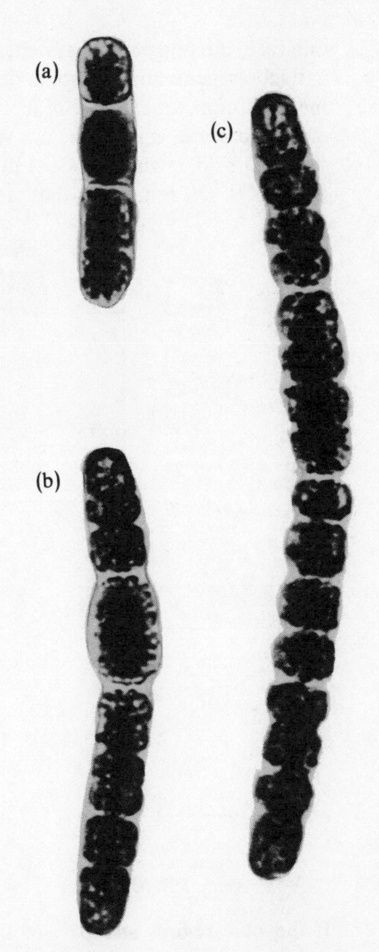

FIGURE 5 (a) A pro-heterocyst in a short filament. (b) and (c) Early and late stages of pro-heterocyst reversion.

2.3 A model of pattern specification

Imagine a row of cells. Suppose (and note that we are only hypothesizing here) that these cells all have the ability to form a pigment of one of three different colours, red, white or blue. (That is, the cells may enter one of three differentiated states.) We want to generate three equal sized regions, one of red cells, one of white cells, and one of blue cells (in other words, a '*French flag*'). How could this be done?

A 'source' of production of some chemical substance might be set up. Such a chemical is a *morphogen* (from the Greek for 'shape' and 'generation'). If the morphogen source is at one end of the row of cells, it will diffuse along the row at a rate depending on the relative molecular mass of its constituent molecules. At the opposite end of the row, there must be a 'sink', which destroys morphogen molecules at a constant rate, so producing a steady concentration drop of morphogen along the row. (If the sink were absent, there would be a danger of morphogen concentration building up to a constant level all along the row.)

If we represent the fall in concentration graphically, we see that each cell in the row has a particular morphogen concentration (Figure 6). To produce regions of red, white and blue, cells might simply be programmed with a specific instruction: for example, if the concentration of morphogen is between T_b and T_w then the 'differentiated state' or colour will be white. This sort of instruction might not be difficult to operate in practice. For example a cellular enzyme may be active only within a certain range of morphogen concentration.

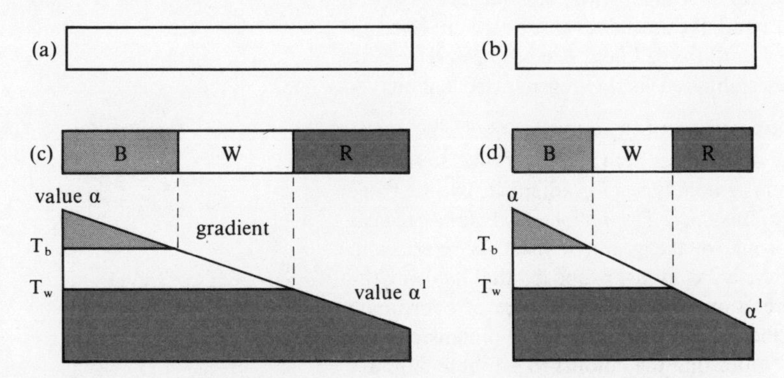

FIGURE 6 The French flag model. A line of cells (a) or a shorter line (b) always undergo differentiation so that one third are blue, one third white and one third red as seen in (c) and (d). This could be achieved if opposite ends of the line of cells have different boundary levels α and α^1 of a diffusible substance so that a gradient is established. Differentiation of cells into blue, white or red depends on a threshold response to the concentration of the substance. If the level is greater than T_b, they differentiate into blue, if below T_w as red, and if the level is between T_b and T_w, cells differentiate as white. Unfortunately, because we are restricted to two colours, our French flag is in grey, white or red. For grey, see blue!

2.3.1 The regeneration of a perturbed pattern

The difference between mosaic and regulative development has already been mentioned, but the regulative process itself may be subdivided. Consider what happens if a gradient of the type outlined in the previous section is 'cut through' and the two halves are separated. Two types of regulative growth are possible. In the first type, *regeneration by remodelling*, there is no growth during regeneration, but the missing parts are remodelled from the remaining stump: the regeneration of *Hydra* (a freshwater polyp—see *A Survey of Living Organisms*) is an example. In

regeneration by remodelling*

this case, the regenerated system is complete but smaller than the original. Alternatively, regeneration in some developing systems is accomplished by *growth* of the cut stump, an example of this being vertebrate limb regeneration. In this type of regeneration, cells at the cut surface are stimulated to undergo extra cell divisions. Consider what happens in each case in terms of our 'French flag' model (Figure 7). Do remember that we are only looking at a model, however.

regeneration by growth*

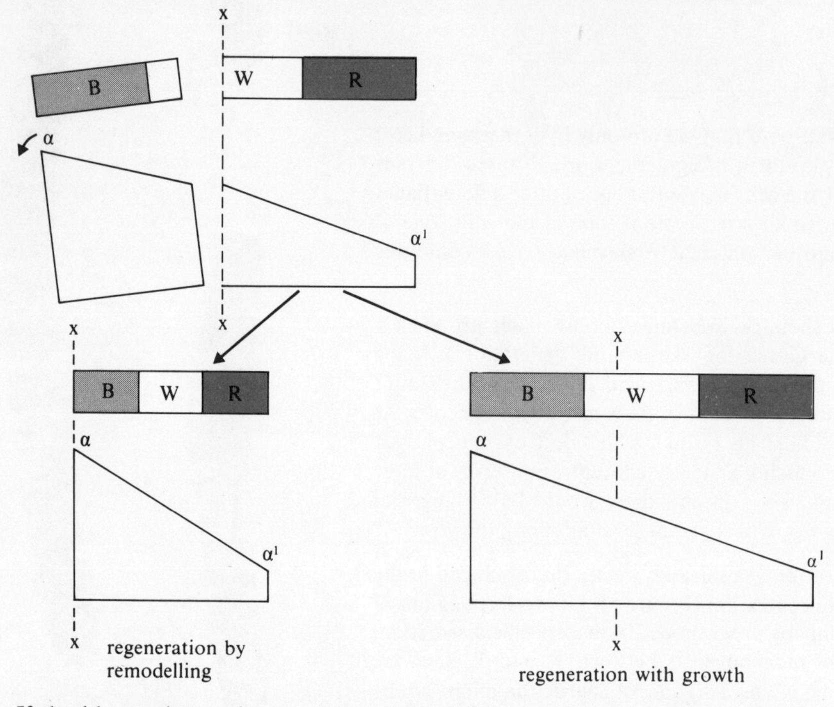

FIGURE 7 Regulation of the French flag. Suppose the pattern is cut at x---x and the left portion removed. In regeneration by remodelling, the value of the positional information at the level of the cut becomes α, and a new, steeper gradient is established. If the cells have the same threshold, a complete but smaller French flag is formed. In regeneration with growth, the boundary value α is established at the end of the newly grown tissue. The slope of the gradient is unchanged.

If the blue region and part of the white region are removed, regeneration by remodelling results in the remaining white and red regions regulating to form the total 'French flag'. The cut surface becomes the new boundary of the field, and the gradient is reset so that the threshold limits of blue, white and red are reset. In regeneration with growth, on the other hand, the gradient value at the cut remains unchanged, and new cells are generated until the full field size has been reformed. The lower gradient values are then re-established in this regenerated region.

One attractive feature of gradients as specifiers of information that cells use to interpret their position in fields during development is that many diverse sorts of pattern could be generated from a very simple type of gradient in two or three dimensions as well as in one dimension. It is the *cell's interpretation of the gradient* which sets up the pattern, not the profile of the gradient itself. Wolpert, who originally formulated the idea of the French flag model proposed that by using the flag model, different rules for interpretation with the same type of coordinate system could give in one dimension the French flag and in two dimensions the Stars and Stripes or the Union Jack. Notice that the colours in all these flags are the same, but that each pattern derives from a particular spatial arrangement of the colours.

2.4 From theory to experiment

Many variations of gradient models have been proposed for setting up patterns during development. The theory behind some of them involves some complex mathematics, but they share certain basic features. What is the experimental evidence for gradients in development? One of the best studied examples is the development of the ripple pattern on the adult *cuticle* of the bloodsucking insect *Rhodnius*. The abdominal segments of this insect bear surface folds that run laterally across each segment parallel to the anterior and posterior segment margins (Figure 8). *Rhodnius* has five larval instars, and a series of experiments may be performed on fifth instar larvae as follows. If a small square of the cuticle is cut out and replaced in the same site in the same orientation, the adult cuticle pattern developing from this individual is normal (Figure 9a). If a square from one segment is swapped with a square from the same position in an adjacent segment, then there is also no alteration in the adult pattern. However, if the cuticle square is replaced after rotating it through 180° (Figure 9b), the pattern of ripples in the adult is disturbed by the formation of local whorls. Also, if anterior and posterior portions of the same segment are exchanged, two whorls of ripples appear.

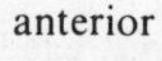

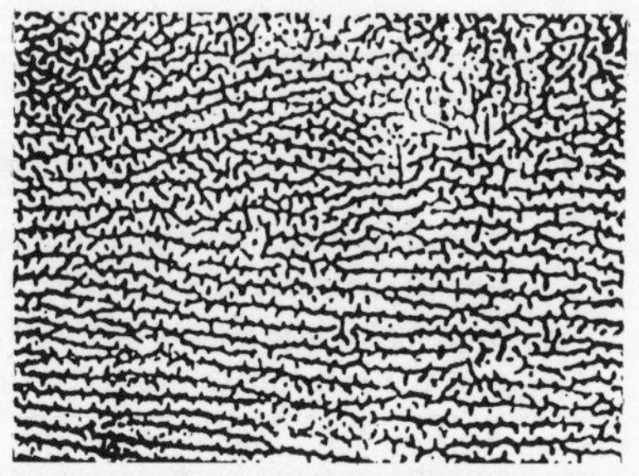

FIGURE 8 Ripples in the cuticle of adult *Rhodnius* (× 150).

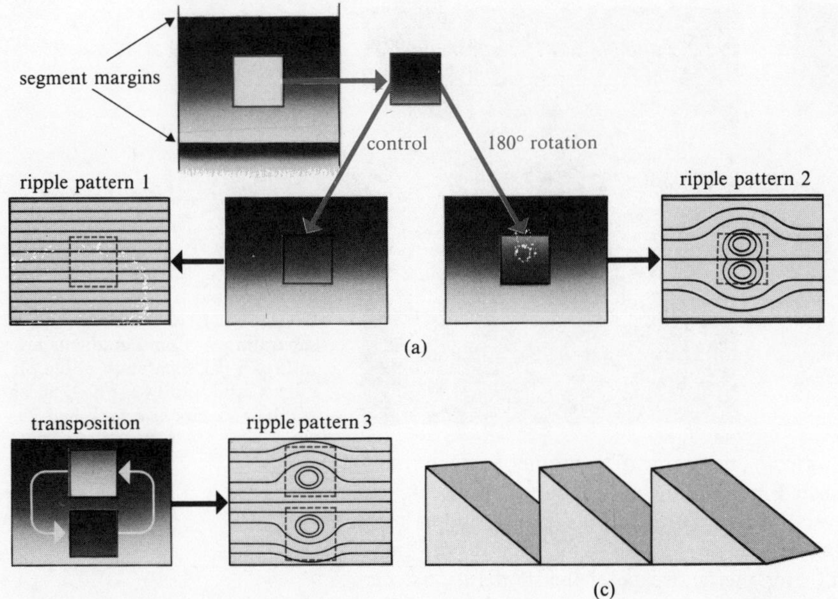

FIGURE 9 (a) The effect of transplantation of pieces of *Rhodnius* cuticle as a control and following 180° rotation. (b) The effect of exchange of anterior and posterior portions of the segment. The gradient in the larval epidermis is represented as a graded grey tone. Where cells at different gradient levels are adjacent, a disturbance of the adult ripple pattern emerges. (c) The serially repeated segmental gradients—the vertical lines represent segment margins.

These experiments were first performed by Michael Locke in the late 1950s. They seemed to indicate some form of communication between the cells of the insect cuticle, as the placing of 'foreign' cells beside one another clearly induced the whorling. What was the underlying mechanism? Locke initially suggested that a biochemical gradient was present along the anterior–posterior axis of each segment (Figure 9c). Cells from different parts of the segment, being differently programmed, were incompatible and therefore their interaction led to a change in ripple orientation.

An important step in demonstrating that the gradient phenomenon might not be restricted to *Rhodnius* came from Lawrence's work on the milkweed bug *Oncopeltus*. The adult bug has a dense mat of antero–posteriorly orientated hairs, and here the hairs act as *polarity* indicators for the underlying epidermal cells. *Oncopeltus* also has a segmental cuticle similar to that of *Rhodnius*, and it is possible to find *Oncopeltus* individuals with gaps in one of their segment margins. The hair pattern around the break is different from normal (Figure 10): it is as if something has pushed through the border and spread out around it. How could a gradient acting across the *Oncopeltus* cuticle segment explain the peculiar behaviour of the hairs around the broken segment border?

polarity*

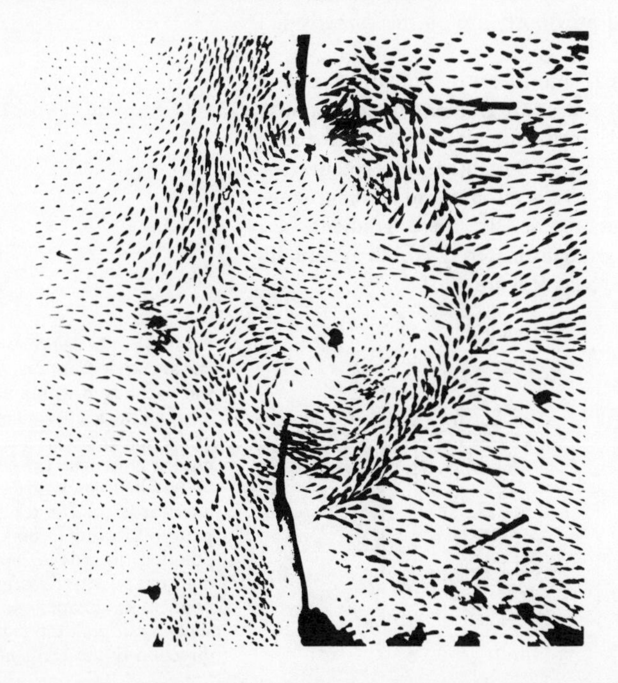

FIGURE 10 The hair pattern surrounding a natural discontinuity in a segment margin of *Oncopeltus*.

Lawrence constructed a sand model to represent the form of a serially repeating gradient. This is shown in Figure 11, where glass plates are used to represent the segment borders, and sand is used as an indicator of gradient level: a stable sand slope between each glass plate represents the gradient. If a break in the glass plate

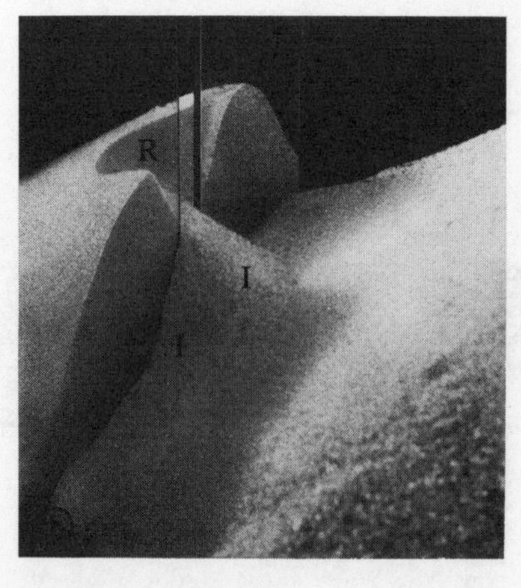

FIGURE 11 Two glass plates separating two sand gradients are shown in (a). In (b), separation of the plates allows sand to flow through and a new stable landscape is established.

develops (see Figure 11b) sand flows from one 'segment' to another and continues until new stable slopes are established. Finally, in the central part of the break, the sand gradient is aligned in the reverse direction (see R, Figure 11b), and in both segments there are gradients of intermediate orientation (see I, Figure 11b). Imagine that lines are now drawn all over the landscape of the model *to indicate the direction of the steepest sand gradients.* Suppose that these lines mark the orientation of *Oncopeltus* hairs, where the hairs point down the steepest sand slopes. Compare Figure 10 with Figure 11b, and note how such a pattern marked on the landscape of the sand model would conform to the hair pattern at the gap in the segment margin.

Because the model fits so neatly, we can postulate that the polarity of the epidermal cells depends on the orientation of the steepest gradient slope. As in the model, when an unstable situation develops as a result of cells that were previously at different gradient levels being placed together (i.e. sand at different levels), a flow of gradient material (sand) occurs until a maximum stable slope is produced. Such an unstable situation occurs at the discontinuity of the segment margin shown in Figure 10 and after the transplantation of pieces of cuticle. Because the flow of gradient material that results alters the gradient levels of all cells affected by it, the direction of the steepest gradients will also change—hence the polarity of the cells is itself altered.

Could these principles also be used to explain the oriented cuticular structures of *Rhodnius*? Analogues of the *Oncopeltus* hairs exist in *Rhodnius*. These are the tubercles (small scale-like cuticular outgrowths) of the ventral abdomen. (Remember that the ripple patterns described previously are on the *dorsal* side.)

Like *Oncopeltus* hairs, these tubercles are polarity indicators that might have their orientation determined by the direction of the steepest gradient slope. Suppose that in the larva, a piece of anterior cuticle from one abdominal segment is exchanged for a piece of posterior cuticle from an adjacent segment (see Figure 12a). Figure 12b shows a hypothetical cross-section of the gradient landscape immediately after the transplantation. Obviously, these gradients are unstable, and as in the sand model, flow of gradient material should eventually produce a stable landscape of maximum steepness (Figure 12d).

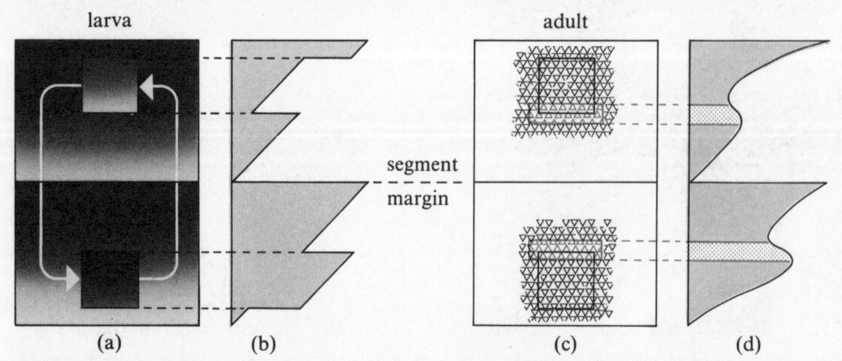

FIGURE 12 (a) An anterior square of larval cuticle from one ventral segment was exchanged with a posterior piece from an adjacent segment. The gradient in the epidermal cells is shown as a graded grey tone. (b) A cross-section of the gradient landscape immediately afterwards. (c) The resulting adult pattern showing in pink the tubercles pointing towards the anterior rather than the posterior margin. (d) The gradient landscape following diffusion, showing in pink the regions where the direction of the gradient is reversed.

Note that the model predicts that in the two regions shown in red the orientation of the tubercles will be reversed, so that they point down the gradient slope towards the anterior margin. Just such a result is evident (see Figure 12c).

Perhaps then the ripples on the dorsal surface of the abdomen of *Rhodnius* are also polarity indicators. If so, their orientation might be determined in much the same way by the direction of the gradient slope. But the ripples run transversely across the segment, so they cannot point down the gradient slope as do *Oncopeltus* hairs. Instead, they must run perpendicular to the lines of the steepest gradient. So the ripples represent contour lines, joining epidermal cells at the same level in the gradient landscape, that is, joining cells that contain identical concentrations of gradient substance (see Figure 13a). A circular whorl of ripples,

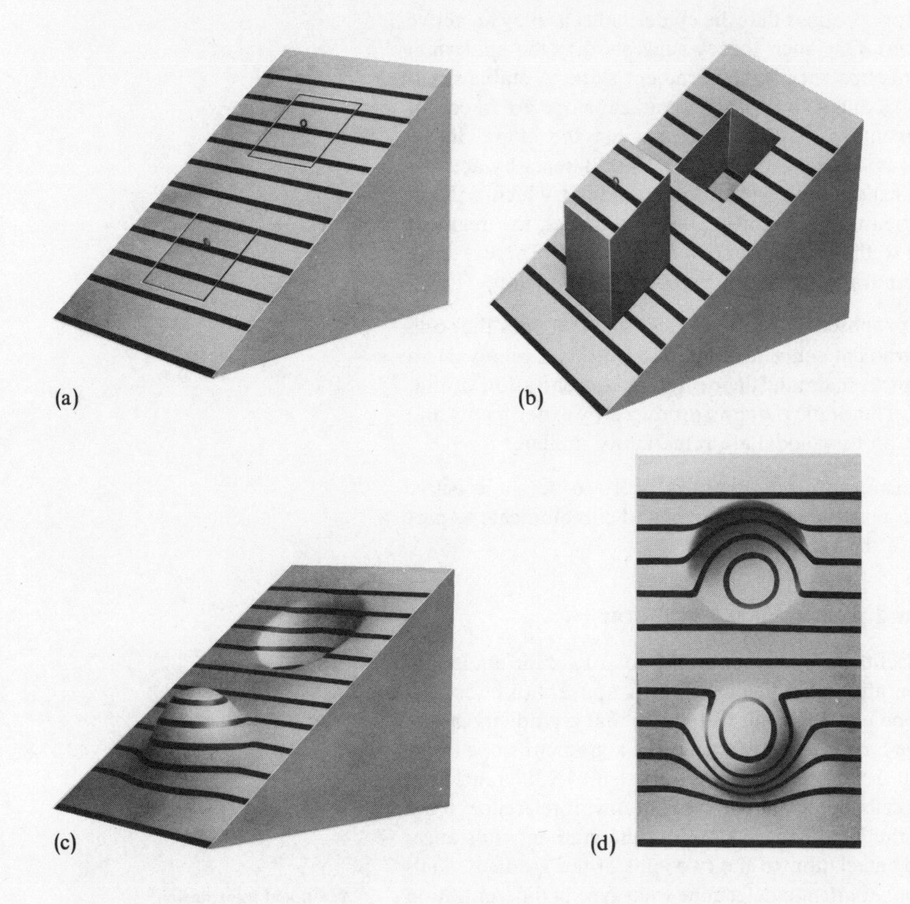

FIGURE 13 The three-dimensional gradient model, showing the effect of transposing anterior and posterior portions on the gradient landscape (a). (b) is before and (c) after diffusion. (d) shows the pattern of contour lines from above.

then, implies that the epidermal cells immediately beneath are all at the same gradient level. If this is so, we should be able to predict the ripple pattern following transplantation by using a three-dimensional gradient model as shown in Figure 13. Consider, for example, Locke's experiment involving an interchange between anterior and posterior portions of a segment (see Figure 9b). Immediately after the exchange of the integument pieces, the gradient landscape resembles Figure 13b. After a flow of gradient material by diffusion, the profile that results is a smooth valley and a rounded hill (see Figure 13c). The contour lines drawn on this three-dimensional landscape, joining points at equal height, form a pattern with two whorls. Viewed from above (Figure 9d), it precisely resembles the ripple pattern that resulted from Locke's experiment (see Figure 9, ripple pattern 3).

Suppose that the gradient is of 'some substance'. How is it established and maintained? A simple source–sink model proposes that one segment margin, say the anterior, produces the gradient substance, and so represents the source, while the posterior margin breaks it down and so represents the sink. Thus, the segment margins maintain two different concentrations of a substance that diffuses passively through the epidermal cells. But consider the following experiment. Adult *Rhodnius* derived from larvae that have been operated on and therefore bear disturbed ripple patterns can be made to form an 'extra' cuticle by injecting moulting hormone (ecdysone). The disturbed pattern in the extra cuticle is very little different from the previous pattern, even though 3–6 weeks elapsed between the deposition of the two cuticles.

□ How is this result incompatible with a gradient model relying solely on diffusion between anterior and posterior margins?

■ If the gradient was maintained by a simple source–sink model, any local disturbances in the gradient would soon be overcome by continued diffusion, and the normal gradient landscape would soon be re-established.

Because the ripple pattern persists, it seems to represent a steady-state landscape, where the forces responsible for maintaining the landscape have come into some sort of equilibrium. It seems that there is some cellular force that can limit the extent of diffusion.

Alternative explanations therefore propose that the epidermal cells play an active part in maintaining the gradient. One such theory suggests that the epidermal cells actively pump the gradient substance up the gradient slope. A stable situation develops where the forces of diffusion down the gradient slope are in equilibrium with the active movement of the substance against the slope. If the direction of the gradient slope is changed, the epidermal cells react by actively transporting the substance against the new concentration gradient, which is therefore maintained in its new orientation. According to this model, the segment margins make no contribution to the maintenance of the gradient. (There is some experimental evidence to the contrary, which we shall not discuss here.)

Another explanation has been proposed by Francis Crick, who imagines that cells can 'remember' their original gradient concentration following their removal to a new site, and that cells attempt to maintain their internal concentration of that substance at the original level. The ripple patterns produced by experiments and by computer simulation based on this model are remarkably similar.

Whatever the precise mechanism, note that there is likely to be some active contribution by the cells, which must therefore interact and communicate as part of a cell population to maintain the gradient.

2.5 The segmental gradient and positional information

The experiments reported in Section 2.4 suggest that the polarity of the epidermal cells, as expressed by the orientation of cuticular features, appears to be locally determined by the steepest slope of a gradient. But notice that a gradient carries two sorts of information. As we have seen, the direction of a gradient slope could indicate polarity. In addition, if the segment margins represent two different, fixed levels of a gradient substance, cells can assess their position with reference to the linear gradient by measuring the local concentration of the gradient substance; the position would be defined in relation to the two ends of the gradient. Cells might then use this *positional information* to determine what type of differentiation is appropriate to their position.

positional information*

For example, the abdominal segments of adult *Galleria*, the wax-moth, bear various sorts of cuticular structures (see Figure 14), while the larval cuticle is more or less homogeneous. Small pieces of cuticle can be cut from various parts of a larval segment and cultured in isolation in the abdominal cavity of a mature caterpillar.

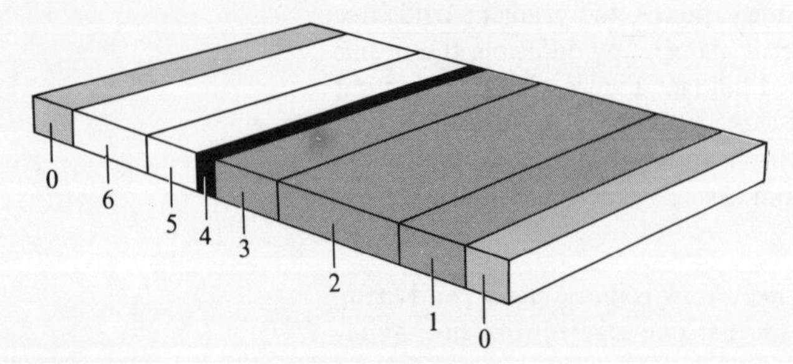

FIGURE 14 Diagrammatic representation of an abdominal segment of adult *Galleria*. The main area of the segment surface bears three kinds of scale, regions 3, 2 and 1 (*grey*). Just anterior to this is a thin ridge of tanned cuticle (region 4, *black*). The segment margin consists of three types of cuticle, 5, 6 and 0. So, each segment margin has a posterior (region 0, *pink*) and an anterior (5 and 6) aspect.

If, after metamorphosis of the host, the pieces of cuticle are removed from the adult moth, they are found to have developed into adult cuticle according to their *presumptive* fate, that is, larval cuticle from presumptive region 3 develops into adult type 3 cuticle. So, the various parts of the segment appear to be determined provisionally before metamorphosis. However, would cells develop according to their prospective fate even when surrounded by cells of a different presumptive type? Fortunately, in *Galleria*, cells from males and females can often be distinguished by characteristics of their nuclei, so when presumptive tissue is transplanted between male and female caterpillars, the cells of the graft and of the host can be distinguished. A small piece of larval cuticle from presumptive region 0 (see Figure 14) was transplanted into larval presumptive region 2. In the adult, all the

presumptive

cells of the graft had produced cuticle according to their origin (type 0), while the host cells (which would normally give rise to type 2 cuticle) produced mainly a ring of type 1 scales around the graft. In addition, the orientation of these host type 1 scales had altered. Instead of pointing posteriorly, they pointed towards the graft.

Try to account for these observations in terms of the segmental gradient before you read on.

Graft region 0 seems to maintain a low gradient position and so causes inflow of gradient material. A local valley is therefore formed in the gradient landscape. As the polarity of the scale-secreting epidermal cells is determined by the steepest gradient slope, this results in the reorientation of the scales. But notice that the developmental fate of some of the host cells within this dip is also changed. The gradient level within presumptive type 2 host cells seems to have been changed by the dip in the landscape to a level corresponding to that of presumptive type 1 cells. This causes presumptive type 2 cells to develop as type 1 cells. So, it appears that it is the position of a cell within the gradient, rather than its position in the segment or its developmental history, that determines its developmental fate.

The important conclusion is that the local level of gradient substance appears to impart 'positional information' to the epidermal cells, which can then interpret and respond to the information by following an appropriate pathway of differentiation.

Summary of Section 2

1 Development may be either regulative allowing repatterning after surgical disruption of the tissue, or mosaic where repatterning does not occur.

2 From the theoretical point of view, a gradient, for example of some kind of biochemical concentration, is the simplest way of specifying a pattern.

3 The French flag model assumes that each cell is assigned a positional value that depends on its position in relation to the two ends of the gradient. Cells presumably interpret these values by switching on the appropriate genes.

4 The French flag model simulates two types of regulative behaviour: regeneration with growth, where the positional value at the cut end is unchanged and the new boundary value is established at the end of the newly grown tissue, and regeneration by remodelling, where the characteristic boundary value is established at the cut end and all the cells within the field are assigned new positional values.

5 The polarity of the epidermal cells of some insects, as expressed in the orientation of cuticular structures, is locally determined by the steepest slope of a segmental gradient.

6 The maintenance of the gradient seems to depend on the active participation of epidermal cells.

7 In *Galleria*, which has different types of scales on the abdominal segment, the type of cuticular structure formed depends on the position of the scale-secreting epidermal cell within the segmental gradient. The orientation of the scales and the sequence of cuticular types in the segment have a common gradient basis.

8 A simple one-dimensional pattern is shown by the alga *Anabaena*, which is differentiated into two types of cells. A chemical gradient of inhibitor substance has been proposed to be important in setting up this pattern.

Objectives and SAQs for Section 2

Now that you have completed this Section, you should be able to:

* describe how a gradient might specify positional information and so help form and maintain a developmental pattern.
* define positional information using the French flag model; discuss how the model applies to two types of regeneration.
* distinguish between mosaic and regulative development.

★ provide a brief account of experimental evidence for the existence of an insect segmental gradient; explain the significance of the sand model, and demonstrate how the gradient might determine cell polarity; predict and interpret the results of experiments involving the transplantation of pieces of insect cuticle.

★ summarize the important features of pattern formation in *Anabaena.*

To test your understanding of this Section, try the following SAQs.

SAQ 1 (*Objectives 2 and 3*) Are the following statements true or false?

(a) The central problem of pattern specification is how the cells in a given embryonic field become differentiated to form a pattern.

(b) The concentration of morphogen in an embryonic field is hypothesized to be at the same level anywhere in the field.

(c) The French flag model does not account for regulative behaviour.

(d) Embryonic development depends on the spatial organization of cellular differentiation.

(e) In regeneration by remodelling, extra cells are produced to replace those removed by surgical manipulations.

(f) A one-dimensional gradient mechanism would satisfactorily generate a striped French flag, but it would be impossible to use it to produce, say, a Union Jack.

SAQ 2 (*Objectives 2–4*) How would the positional value of cells be established within (a) a regenerating leg stump (regeneration with growth) and (b) a regenerating *Hydra* (regeneration by remodelling)? Assume the presence of a biochemical gradient.

SAQ 3 (*Objectives 3 and 5*) Which of the experimental procedures (a)–(g) illustrated in Figure 15 would result in a distortion of the ripple pattern in *Rhodnius*? In each case, A marks the site from which the graft was removed, and B its final location. If the graft was rotated, the extent of rotation is indicated.

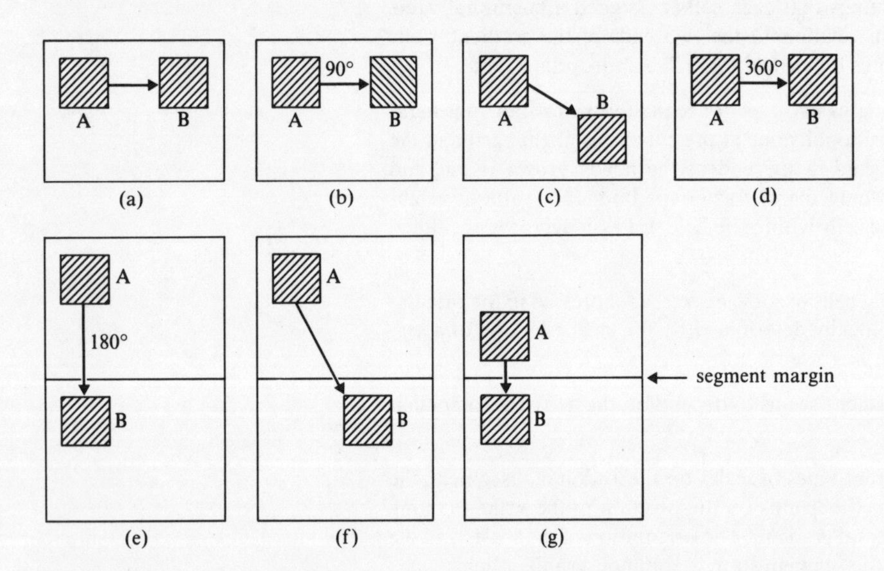

FIGURE 15

SAQ 4 (*Objectives 2 and 5*) The adult segment of *Galleria* is divided into strips of cuticles of seven different kinds (see Figure 14). The main area of the segment bears three different types of scales (regions 3, 2, 1). Just in front of the scales, there is a thickened ridge (region 4). The segment margin consists of three types of cuticle, regions 5, 6 and 0. Consider two abdominal segments A and B, where A is in front of B. The segment margin between these adjacent segments has one part (region 0 in pink in Figure 14) that can be considered as part of the posterior portion of segment A, that is, it is the posterior part of the segment margin. The other part of the margin (regions 2 and 1) can be considered as part of the anterior portion of segment B, that is, it is the anterior part of the segment margin.

In an experiment on larval *Galleria*, a graft containing presumptive regions 0 and 1 was implanted into presumptive region 2 of host larval cuticle as shown in Figure 16a (note the graft therefore contains both posterior and anterior parts of

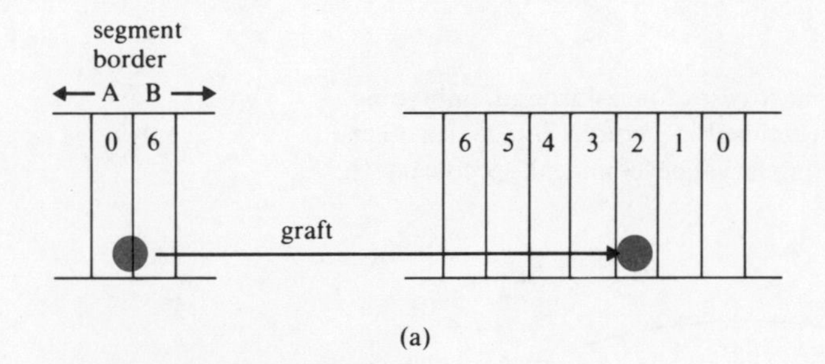

(a)

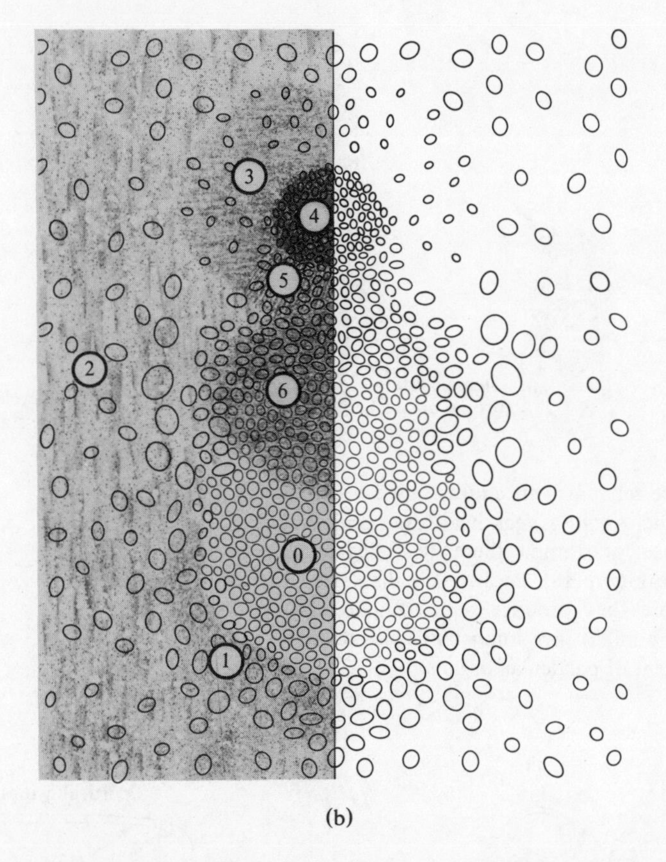

(b)

FIGURE 16

the segment margin). Figure 16b shows the effect of the implant on the cuticular structures that later developed in the host. To the left of Figure 16b, the various types of cuticular structures are shown and identified as types 0–6. Throughout the Figure, the small circles represent the cell nuclei in the epidermis underlying the cuticle. Nuclei of *graft* cells are shown as red; those of host cells as black.

How can the results in Figure 16 be interpreted in terms of a 'gradient of positional information' in the insect segment that determines the type of cuticular structure that epidermal cells produce?

3 *Drosophila* and development

Having looked at some isolated examples of patterns, we now look at the development of a particular organism, the fruitfly *Drosophila*, and study pattern formation in this insect. *Drosophila* has become perhaps *the* most studied developmental organism in the past 10 years, partly because of its interesting and relatively simple development compared with vertebrates, partly because of its ease of rearing and rapid life cycle, and also because of the vast body of genetic studies that have been done on *Drosophila*. There is a huge catalogue of mutations affecting various developmental events, and a good way of trying to understand the *genetic control* of development is to study such mutants. This aspect will be dealt with in detail in Section 9.

See *The S202 Picture Book* for *Drosophila melanogaster*

Drosophila

Drosophila also has an amount of DNA (120×10^6 base pairs) midway between that of *E. coli* (6×10^6 base pairs) and mammals ($2\,800 \times 10^6$ base pairs). This suggests that its genetic complexity may also be somewhere between bacteria and the higher vertebrates.

3.1 The biology of *Drosophila*

Although the study of morphogenesis in most cases hinges around embryonic development, the life history of *Drosophila* (outlined in Figure 17) includes other developmental events open to analysis. During larval development, specific pools

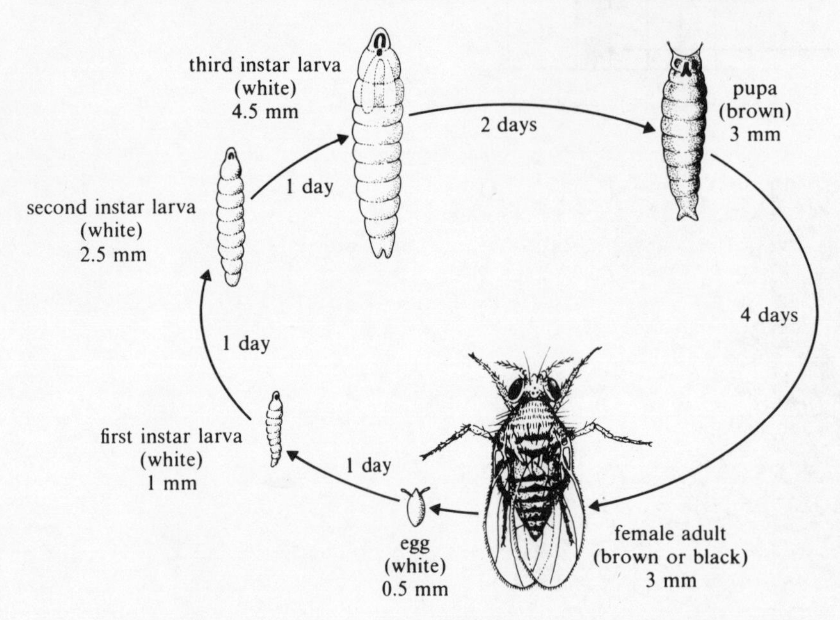

FIGURE 17 Life cycle of *Drosophila*. The durations of the stages are for a temperature of 25 °C.

of cells are set aside in the larva, and these cells form structures called *imaginal discs*. All the adult epidermal derivatives, such as wings, legs, head, eyes, genitalia, and so on, derive from specific imaginal discs; for example, there are six 'leg discs', each of which produces one adult leg (see Figure 18). Because the discs are so relatively simple in structure, and because they produce such complex end-products, it was thought possible that they might give important clues to how pattern formation occurs in the development of particular organs.

imaginal disc*

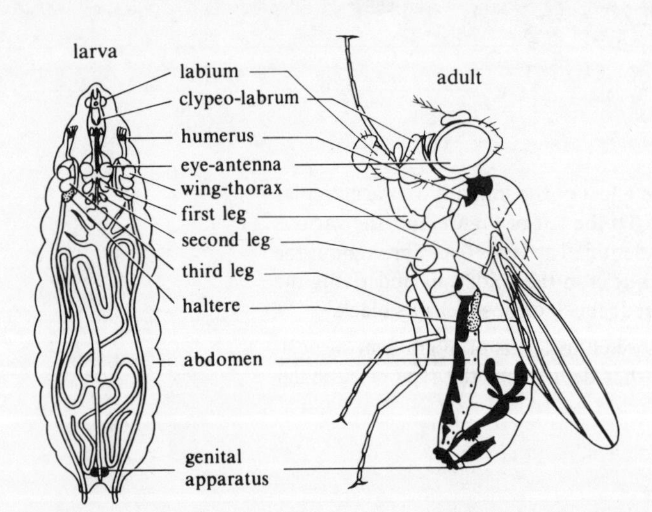

FIGURE 18 Schematic representation of the larval organization and the location of the different discs. Discs and their corresponding adult derivatives are connected by lines.

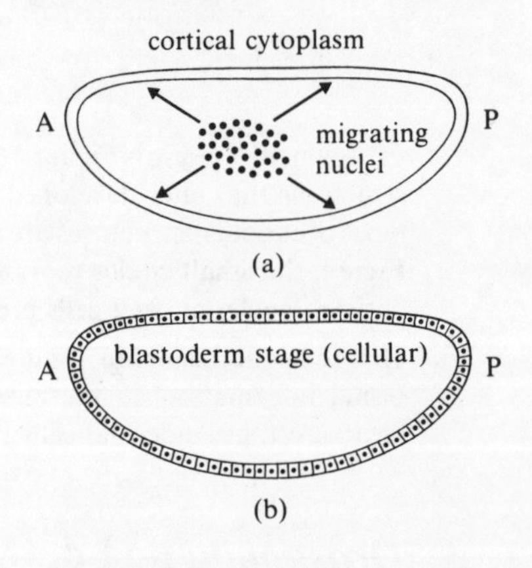

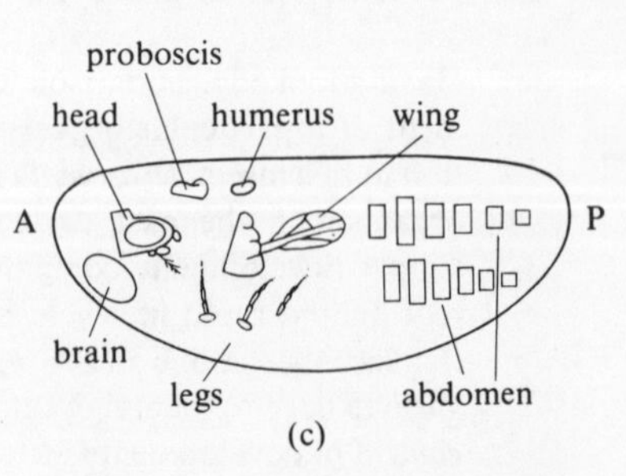

FIGURE 19 Early cell determination in the *Drosophila* embryo. Early nuclear division (a) and subsequent migration of nuclei to the egg cortex (b). Preliminary localization of the precursors of various structures (c).

Let us look at the embryonic development of *Drosophila*, and see how a typical disc is formed. After fertilization of the egg, the diploid nucleus begins to divide, going through enough divisions to produce about 4000 nuclei in the centre of the *Drosophila* egg. After several hours the nuclei migrate out to the periphery of the egg and embed themselves in the *cortical cytoplasm* there. The next step is the acquisition and parcelling up of cytoplasm by the nuclei from the surrounding cortex to form cells. This is called the *blastoderm* stage, and it seems to be here that the cells first 'decide' in broad terms on their subsequent fate (Figure 19).

It is possible to work out the exact stage of development at which cells become determined to form particular adult structures. In other words, it is possible to establish when 'undetermined' precursor cells form subgroups of 'wing' or 'leg' cells and so on. *Drosophila* development is highly mosaic, and this may be shown by pricking eggs with a sharp needle—they then develop with defects in the adult corresponding to the pricked region in the egg.

□ What would you expect regulative eggs to do if pricked?

■ Regeneration would replace the damaged cells, and no defect would be seen in the adult.

By the beginning of larval development, each leg disc consists of a small nest of 20 or so cells, which will increase in number to about 30 000 by the onset of *metamorphosis.* It has been found that determination is a continuous process, because 'leg' cells will subsequently become sub-determined into tibia, femur, etc. The various stages in the development of a leg disc are indicated in Figure 20.

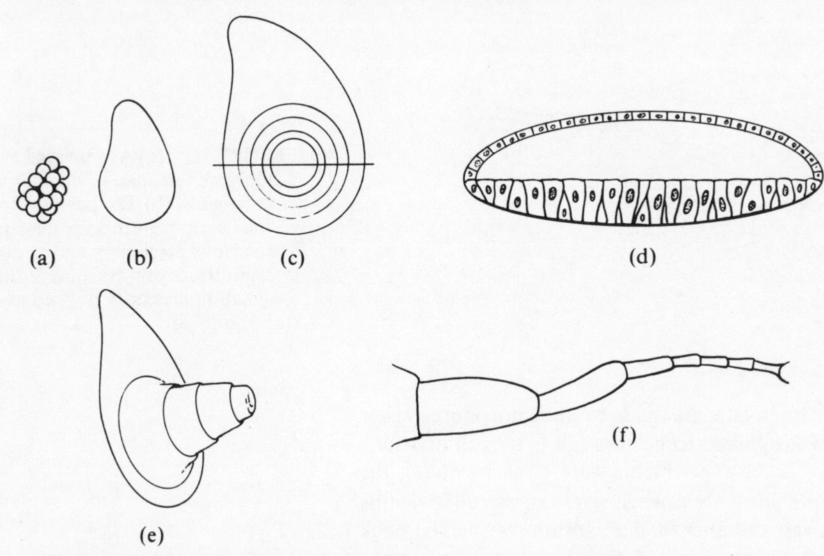

FIGURE 20 Stages in leg disc development. (a) A nest of cells at the blastoderm stage give rise to a disc primordium (b), which grows to a mature disc (c) at the end of larval development. The disc is a single-layered epithelium as in (d). Metamorphosis during pupation gives evagination of leg structures (e), resulting in adult leg (f).

As the four day period of larval growth proceeds, the disc grows in size and takes on its characteristic shape. It is only during metamorphosis, however, that the actual leg structure begins to become apparent. How are the cells determined to become specific parts of the adult leg? Why do they differentiate to give certain structures?

3.2 Imaginal disc transplantation

Like other insects, *Drosophila* has an open circulatory system with blood filling the body haemocoel (see Units 1, 2 and 19). In the 1930s, Beadle and Tatum discovered that it was possible to remove imaginal discs from the larva and to place them either in the body cavities of other larvae or else in the abdomens of adult flies. Because of the nutritious blood acting as a kind of tissue culture medium, such transplanted discs will grow quite successfully.

Before discs differentiate into adult structures, they must be stimulated by the hormones normally present in the blood at metamorphosis. If discs are transplanted from mature larvae into adult abdomens, they undergo cell proliferation, that is, cells divide and die, but they will *not* differentiate. If the discs are transplanted back to mature larvae that are then allowed to metamorphose, the transplanted discs also metamorphose, and the adult structures produced may be dissected out of the body cavities of the host flies.

These techniques have been used to investigate two main phenomena, competence of discs, and regeneration and duplication of disc halves, which are further discussed in the TV programme *Patterns in Development: Gradients.*

3.2.1 Competence of discs

First, transplants can be used to investigate both the *time* and the *spatial pattern* of disc determination. If discs from immature larvae of various ages are transplanted into larvae that are about to metamorphose, then the ability of discs at particular ages to differentiate, or their *competence*, can be studied. The mature larva immediately metamorphoses, making the implanted immature disc metamorphose along with it. Experiments showed that young discs transplanted at the

competence*

mid-larval stage into metamorphosing larvae form adult tissues that have limited features of the normal adult tissues formed from the discs. Progressively older discs can form more and more of the normal adult structures on transplantation. So, it seems that the determination of discs proceeds in steps.

The spatial pattern of determination can be worked out by transplanting fragments of discs of relevant age. After metamorphosis, only the adult structures derived from these parts of the disc will be formed. Thus *fate maps* have been constructed for all the various discs (see Figure 21).

fate maps*

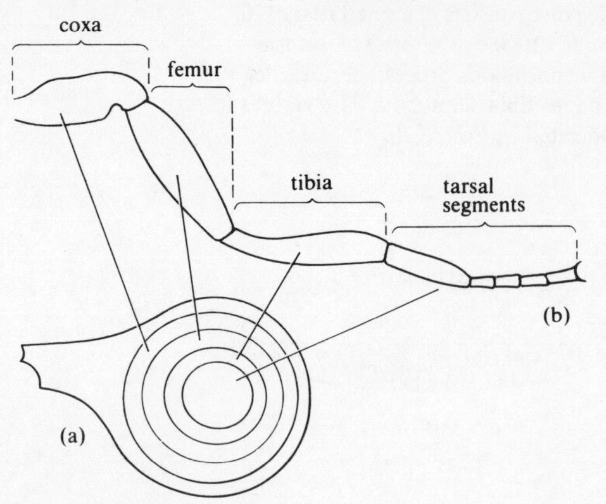

FIGURE 21 (a) A simplified version of Schubiger's fate map of the first leg disc of *Drosophila.* (b) The positions of the various main segments on the adult leg. The various precursors on the disc have a concentric layout because of the evagination necessary to produce the leg (see Figure 20).

3.2.2 Regeneration and duplication of disc halves

We have seen what happens if disc fragments are made to metamorphose earlier than normal. What happens if they are forced to go through *extra* cell divisions before metamorphosis? To find out, some experiments were done involving the transplantation of disc fragments into adult abdomens, where extra cell divisions occur. Each disc is cut into two halves, and after further culture transferred back to larvae and allowed to metamorphose. From fate maps, we know what structures should be formed by a particular disc fragment. What happens when extra time is spent in the adult abdomen?

The results of this sort of experiment show that after metamorphosis of the larva there are two possible results, depending on which of the two disc fragments is studied. One fragment gives *regenerated* parts of the structures normally formed from the disc fragment cut away in the adult. The complementary fragment gives a *mirror image duplicate* in the adult of the cultured fragment. These experiments were carried out extensively by Schubiger in 1971, and he noticed the following points. (*Refer to Figure 22 for details of orientation.*) The upper part of a bisected leg disc (which normally forms the dorsal side of the adult leg) can *regenerate* the lower part (ventral in the adult). However, the complementary lower part of the disc undergoes *duplication.* Also, the medial (inner) half of the disc (see Figure 22)

mirror image duplicate*

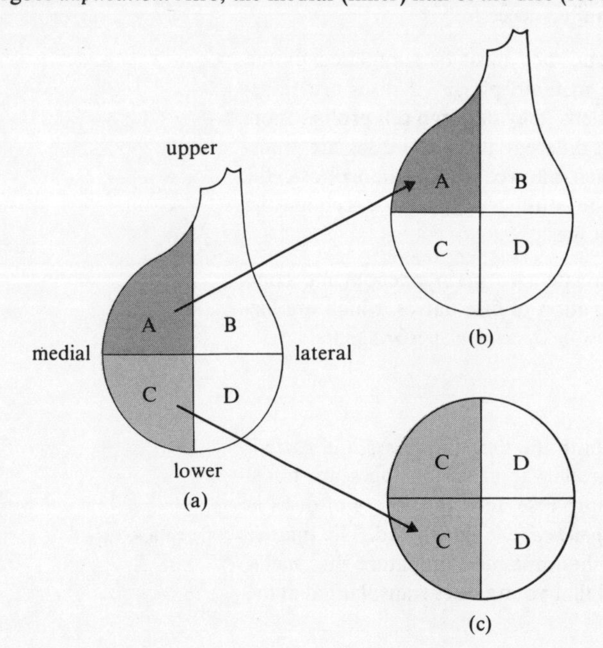

FIGURE 22 Results of a regeneration/duplication experiment in *Drosophila.* The disc in (a) was fragmented and regions A and C allowed to undergo cell divisions by being cultured in adult abdomens before going through metamorphosis. (b) Region A can regenerate all the structures normally formed by the disc. (c) Region C can only duplicate itself and re-form certain limited regions.

can regenerate the lateral (outer) half, whereas the lateral half duplicates. The upper medial quarter of the disc can regenerate a whole leg!

Surgical bisection of the disc results in the formation of a *blastema* at each cut surface, and these blastemas give rise to the regenerate and the duplicate. A blastema is the mass of presumably *de-differentiated* cells (i.e. cells that have lost their differentiated state). It is formed at the cut surface of a regenerating system before re-differentiation and regulation.

blastema

de-differentiation

Serial culturing—transfer of the discs to a younger adult host when the previous host is old—may be performed (Figure 23). If this is done for many generations, a new phenomenon emerges when the discs are finally transferred back to larvae to

serial culturing

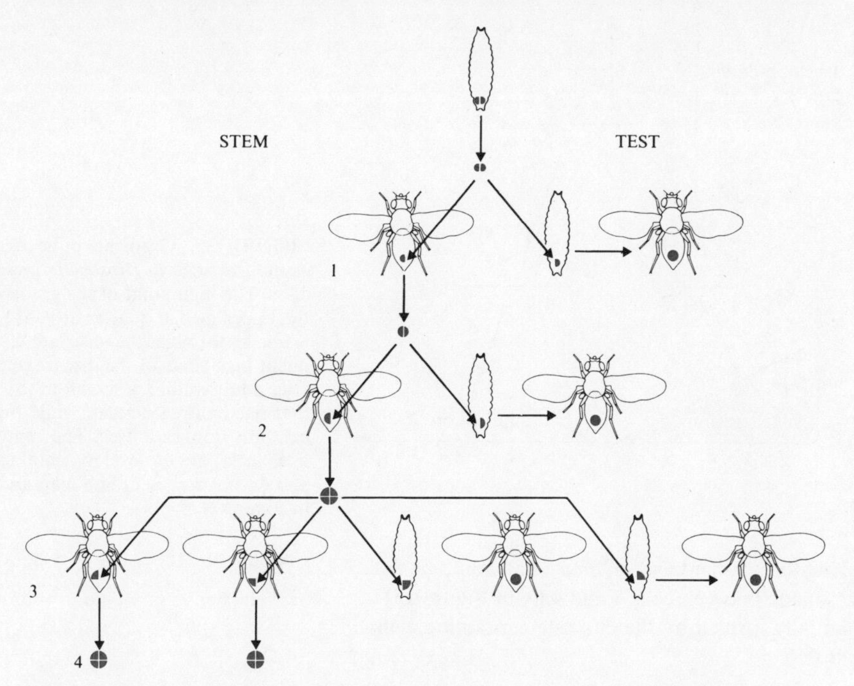

FIGURE 23 Serial transfers of cells grown from a single bit of imaginal disc are shown over four generations. A bit of genital disc from a larva is divided between an adult host (left) and a larva (right), which is allowed to mature so that the condition of the transferred cells can be checked. The process is repeated (2) by retransplanting the disc cells of the original host. By the next transplant (3), the disc cells have increased sufficiently to provide two transplants for testing and two transplants into stem hosts. In the next generation (4), the number quadruples. Disc cells have survived and multiplied in this way for more than 150 transfer generations.

allow metamorphosis to take place. It is found that in such instances, a few discs have undergone *transdetermination*, a switching from one determined state to another. These changes may be very striking, for example, a wing disc before serial transfer may give genitalia or leg tissue after, say, 20 transfers from adult abdomen to adult abdomen. Although the basis of transdetermination is unclear, a further 'clue' is provided by the strict rules that govern how it occurs. Although you are not expected to remember these rules, they are illustrated in Figure 24. Certain

transdetermination*

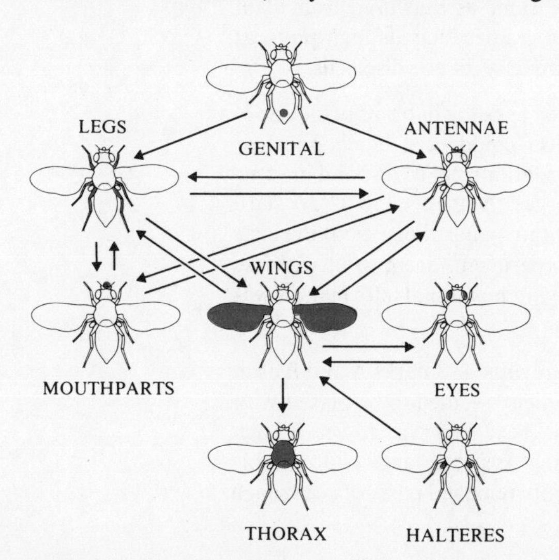

FIGURE 24 The transdetermination sequence undergone by seven kinds of imaginal disc cells is shown by arrows. Genital cells, for example, may only change into leg or antenna cells, whereas leg and antenna cells may become mouthpart or wing cells. In most instances, the final transdetermination is from wing cell to thorax cell: the change to thorax appears to be irreversible.

cell types can be formed interchangeably, while others represent 'terminal' determinations which do not allow subsequent transdetermination: for example, 'leg' and 'wing' are interchangeable, while thorax gives only thorax tissue. A possible explanation is that transdetermination represents a switching of determined states at the genetic level: perhaps single genes can control whether a cell expresses 'thorax' properties or 'leg' properties. We will return to this question when we look at mutations affecting determination in Section 9.4.

3.3 Gradients in imaginal discs

Bryant suggested that the regeneration/duplication experiments could be explained in terms of a gradient of developmental capacity within the disc. The mechanism is explained diagrammatically in Figure 25. Bryant proposed that only disc parts from *higher up* the gradient may form those *lower down*. Lower parts can only form their own kind. For several years, this simple model was used to explain the results of the regeneration and duplication experiments, increasingly fine cutting methods being used to gauge which parts of the discs would regenerate and which would duplicate.

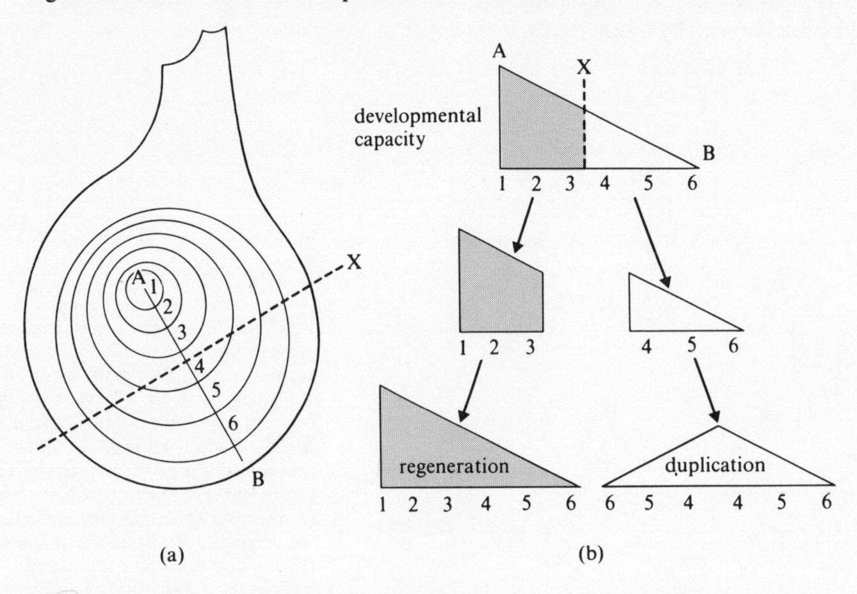

FIGURE 25 Gradients of developmental capacity in *Drosophila* imaginal discs. The high point of the gradient is in the upper medial quarter at A. (a) A cut is made through the disc at X. Subsequent behaviour of the disc fragments after adult culture is shown in (b). Note that one half regenerates, while the other can only duplicate itself. The contour lines in (a) are of developmental capacity and do not represent fate map areas as in Figure 21.

If we specify six particular cells along the gradient as 1–6, then what would be the theoretical expectation if a cut was made between cells 3 and 4 (X in Figure 25)? Two complementary halves are initially formed by the cut, one containing cells 1–3, and the other containing cells 4–6.

□ What cell types could be formed from the 1–3 fragment, and what from the 4–6 fragment after blastema formation?

■ Because the 1–3 cells are high in developmental capacity, they can regenerate the lower cells 4–6. The 4–6 fragment is unable to regenerate *up* the gradient however, and so it duplicates 4–6 cells.

The gradient shown in Figure 25 is basically the same as that used to explain *Rhodnius* and *Anabaena* pattern specification. The gradient has its high point at the left, and is measured in the 'developmental capacity' of the disc cells.

Although the gradient has been cunningly described as being in terms of 'developmental capacity', its actual physical manifestation could be a biochemical or electrical gradient of some kind. Cells would be 'programmed' to respond in some way to specific concentrations or electrical charges. As yet, there is no hard physical evidence for the existence of such a gradient in imaginal discs, but the facts gleaned from the regeneration and duplication experiments seem to fit with the results that would be expected if a gradient specifying positional information was present.

Figure 25b shows an impression of the disc gradient with landmarks A and B as in Figure 25a. The experiments showed that the 'gradient' seemed to be placed with its high point in the upper medial quarter. The next step was to see if such gradients could be located in other regenerating systems, and two notable examples were found in the regeneration of surgically removed parts of cockroach legs and of parts of amphibian limbs.

3.4 Models and development

It is arguable how closely these 'wound healing' results can be applied to what goes on during normal development. However, the majority of examples of pattern formation are so complicated that it is much easier to study simpler models of these systems. A system with certain attributes of a second system may therefore be used as a model of that second system. Modelling is very important in the study of complex developing systems. The use of models in developmental biology was discussed in the TV programme introducing the development Block, *What is Development*? For instance, pattern specification in *Anabaena* might be considered a model for more complex development in higher organisms. In both cases, we have the same basic problem—how do cells acquire particular 'positional information', and how do they decide what sort of cell to differentiate into.

Simple organisms often have general properties present in more complex animals or plants. The usefulness of regeneration studies to our understanding of normal development is more problematical. Although regeneration involves many processes seen in normal development (cell division, differentiation, morphogenesis), it involves abnormal development—cutting discs up, transplanting them to adult abdomens to grow, and transplanting them back again to larvae to metamorphose. Does this really prove that the gradient is acting in normal morphogenesis? We will leave you to ponder on these problems—there is no clear solution at present.

Summary of Section 3

1 *Drosophila* is useful for developmental studies because development is rapid, and amenable to experimental analysis.

2 The imaginal discs, larval precursors of adult epidermal structures, are two-dimensional sheets of cells used to study pattern formation and morphogenesis.

3 Imaginal discs or parts of discs may be transplanted from larvae to the abdomens of adult flies where they will grow but not differentiate. Transplantation back from adult to larva allows metamorphosis of the discs along with their larval hosts.

4 Transplantation can be used to observe disc determination by observing the behaviour of disc fragments transplanted at different times of larval development into other larvae about to metamorphose. It can also give valuable insight into the mechanism of pattern specification in discs as shown by regeneration/duplication experiments involving transplantation of disc fragments into the adult abdomen.

5 It has been proposed that a gradient of pattern-forming ability exists in the cells of imaginal discs.

6 Simple developing systems may be used as models of more complex systems.

Objectives and SAQs for Section 3

Now that you have completed this Section, you should be able to:

★ briefly describe the life cycle of *Drosophila*, and give reasons for the usefulness of *Drosophila* as an organism for developmental studies. Discuss why the structure and biology of imaginal discs make them useful models for pattern formation.

★ describe how regeneration and duplication experiments in *Drosophila* imaginal discs give evidence for a gradient of positional information.

To test your understanding of this Section, try the following SAQs.

SAQ 5 (*Objectives 7 and 8*) An experimenter who wants to test the gradient model of imaginal disc regeneration/duplication takes a slice of tissue containing regions 4 and 5 (as in Figure 25a). What tissue types would this fragment give after culturing in an adult and subsequent metamorphosis?

SAQ 6 (*Objectives 7 and 8*) The following surgical operations were performed on mature imaginal leg discs after removal from larvae and before transplantation to adult abdomens to allow regenerative cell divisions (Figure 26). In (a), fragments W and X were separated, and in (b), fragments Y and Z. After a period of growth in the adult abdomens and subsequent transfer back to a larval host to allow metamorphosis, which of W–Z showed regeneration and which showed mirror image duplication?

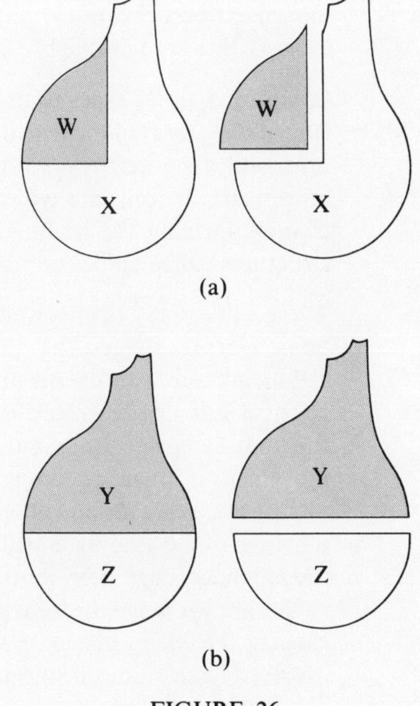

FIGURE 26

SAQ 7 (*Objectives 7 and 8*) Which descriptions of the following terms are correct?

(a) Transdetermination is:

(i) The process of cell division in insect larvae.

(ii) The change in state of previously determined cells.

(iii) The process by which determined cells lose the capacity to divide.

(iv) The process by which groups of cells formed in the larval stage of insects are determined to be specific parts of the adult insect.

(b) Cortical cytoplasm is:

(i) Maternal cytoplasm.

(ii) The layer of cytoplasm at the egg periphery to which cleavage nuclei migrate.

(iii) Cytoplasm found in imaginal discs.

(iv) Cytoplasm found in the centre of the egg.

(c) Competence is:

(i) A stimulus emitted from a group of cells.

(ii) The name given to the process of differentiation.

(iii) The relative ability of cells to differentiate.

(iv) The process of cell proliferation during development.

4 Cells and morphogenesis

Until now, we have been looking at the problem of how patterns are set up, and have used the gradient as a basic specifying mechanism. How do cells use the information they get from pattern-specifying mechanisms to build shapes? The study of the shape-building aspects of morphogenesis roughly splits into two parts. First, the study of the properties of the individual cells. This seems a promising line of attack, because if we have some idea of how the individual units of development, the cells, work, then this might give us an insight into how cells cooperate to build structures. We concentrate on this approach in this short Section. Second, we will go on to look at how cells cooperate during morphogenesis—this is taken up in Sections 5–7. The TV programme Patterns in Development: Cell Movement *illustrates many of the principles outlined in Sections 4–7.*

If we take an embryo, separate the constituent cells chemically and plate the cells out in suitable medium in a Petri dish, the cells do not just remain static, but move around. There are two types of movement. Some cells (called *amoebocytes*) move, as their name suggests, rather like amoebae: directional movement is associated with extension of newly formed pseudopodia at the leading edge and a compensating retraction of cytoplasm from the trailing edge (Figure 27). Exactly how this movement occurs is not known, but cytoplasmic streaming is an obvious feature. Cells of this type are highly motile in the adult tissue *in vivo*.

FIGURE 27 Outline of a lymphocyte in tissue culture, showing amoeboid movement.

In contrast, other types of tissue cells, for example *fibroblasts*, show a characteristic gliding form of movement. Such cells adhere to a solid or semi-solid substratum, and glide across it with no obvious pseudopodial formation. During this movement, the cells are typically triangular in shape, with the broad base of the triangle forming the leading edge or *leading lamella*. The *leading lamella* bears structures called ruffles on its edge (see Figure 28).

These ruffles may be involved in cell movement, but how they work is not clear. Some workers suggest that the ruffles are evidence of intermittent contact of the cell membrane with the substratum. The continuing undulating movement of the edge of the leading lamella may produce waves of adhesive contact with the underlying substratum, thus moving the cell forward. Time lapse film of cell movement in tissue culture is shown in the TV programme *Cell Movements*. An alternative explanation is that cell movement involves the forward extension of the edge of the leading lamella and subsequent attachment to the substratum. If the extended edge now contracts, the whole cell is drawn forward. In this case, ruffles appear when the front edge fails to attach successfully to the substratum, so causing the edge to bend upwards and backwards. So, rather than being directly involved in cell movement, ruffles might indicate a periodic failure of the cell and substratum to adhere.

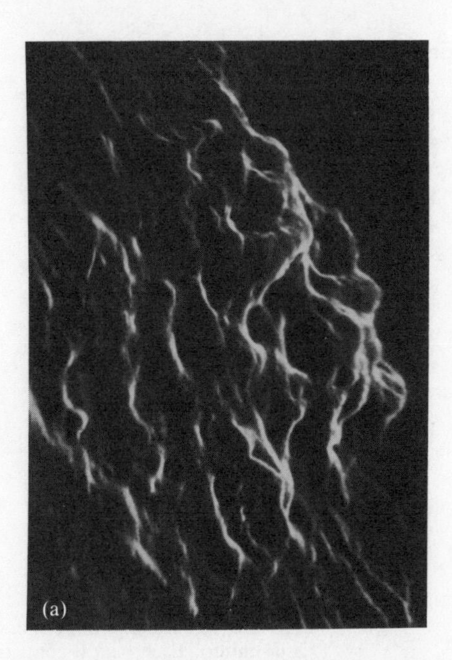

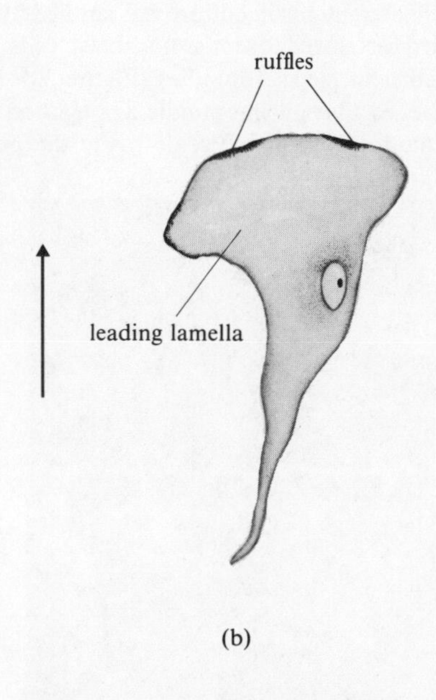

FIGURE 28 A fibroblast from a chick heart. (a) A scanning electron microscope picture of an instantly frozen cell shows the ruffled membrane structure. (×5 700) (b) A fibroblast moving over the substratum in the direction of the arrow.

cell alignment

Not only cell movement, but also *cell alignment* under culture conditions can give information about cell behaviour. Paul Weiss, a pioneer in this area, discovered that fibroblast cells from embryonic chick hearts oriented themselves on fine parallel grooves cut into a glass plate. This suggested that cell movements might be oriented by special characteristics of the substratum over which they move. Weiss termed this phenomenon *contact guidance*, and thought it might be explained by the secretion of a macromolecular exudate laid down on the substratum by the migrating cells. According to Weiss, the molecular network of such a mat could be oriented by the groove. Despite a later discovery that cells in culture do release macromolecular materials that coat the substratum, Weiss's theory is disputed. For instance, as more surface is available within grooves than in the flat areas between them, there may be a greater probability of contact for leading lamellae in the grooves. Thus contact guidance might not depend on special orienting properties of secreted materials but purely on the contact between groove and cells. Alternatively, the adhesive characteristics of the surface within the groove might differ from those of the plane surface outside. In each instance, the cell would become trapped in the grooves.

contact guidance*

The alignment of cells in culture can also give information on the mechanism of differentiation. The American worker Konigsberg discovered around 1960 that if a suspension of chemically disaggregated cells from chick muscle are placed in a suitable medium in a plastic Petri dish, the cells de-differentiate to form fibroblast cells that cling to the substratum. If a layer of the protein collagen is put on the Petri plate before the experiment, the cells become fused together and eventually differentiate into whole muscle tubes (Figure 29). Clearly, the interaction between substrate and cells is of prime importance here.

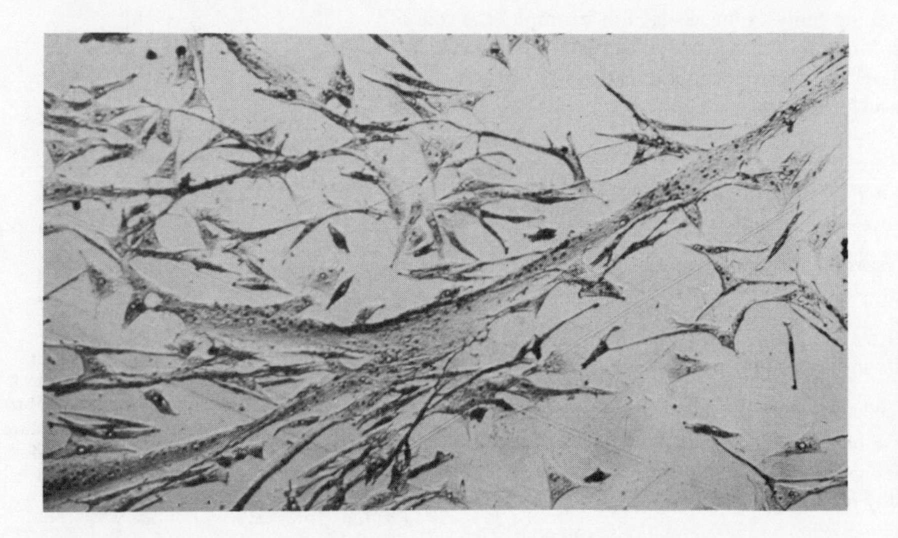

FIGURE 29 Muscle tube fusion *in vitro*. Fibroblast-like cells aggregate together to form parallel bundles and differentiation into myotubes takes place. Under a microscope, the distinctive banding pattern of the myotubes may be seen.

Another interesting example of cell interactions in culture was studied by Elsdale, who was interested in how, under certain conditions, fibroblast cells in tissue culture can align themselves together in parallel bundles (Figure 30). By using time-lapse cinematography, he noticed that during bundle aggregation the cells underwent a kind of see-sawing motion together. Elsdale made the novel suggestion that the fibroblast bundle was behaving as an example of what is called in engineering an *inherently precise machine*. One such machine grinds spherical lenses by randomly rubbing two blanks over one another as shown in Figure 31. Similar random movements (the 'see-sawing' observed by Elsdale) could be acting in the fibroblast bundles, only instead of forming a 'spherical' shape, the nature of the cellular material forces the cells into a parallel bundle. This idea is highly speculative, but it raises the possibility that large groups of cells, all doing identical simple things (like 'see-sawing'), can undergo a primitive form of morphogenesis.

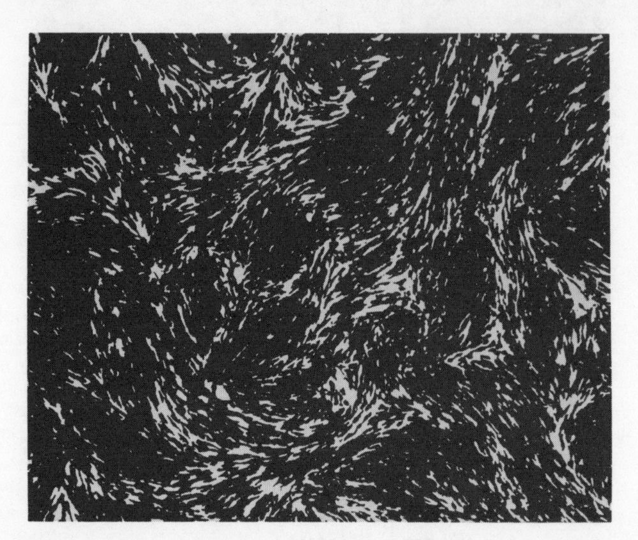

FIGURE 30 Bundles of fibroblasts in tissue culture. Each bundle consists of many individual spindle-shaped cells arranged in parallel fashion.

inherently precise machine

A final property of cells *in vitro* to be mentioned here is that of *contact inhibition*. When one fibroblast contacts another in cell culture, the advancing cell contracts somewhat, its leading lamella is paralysed, and movement stops. The effect may be either physical or chemical—for example, a localized accumulation of acid metabolites between two opposed cell surfaces. Possibly, contact inhibition is specific to the cell type and therefore may be of considerable importance *in vivo*. For example, invasive cancer cells are not inhibited by contact with fibroblasts in culture. But in contrast, normal cell lines from different origins do tend to show mutual contact inhibition, so its precise significance is in doubt. In the next Section and in the TV programme *Cell Movements*, we see how cells can sort themselves out into specific types, and possibly contact inhibition has a role to play here.

contact inhibition

What relevance do these various *in vitro* phenomena have for normal morphogenesis? The *in vivo* behaviour of cells in development is hard to analyse because of microscopic limitation, although technical advances such as scanning electron microscopy (see Unit 4) are making things a bit easier. For example, the role of contact guidance and contact inhibition in early embryology is not known. At the most, these *in vitro* studies can indicate several general features of cell behaviour that might form a tentative basis for understanding the behaviour of cells during morphogenesis.

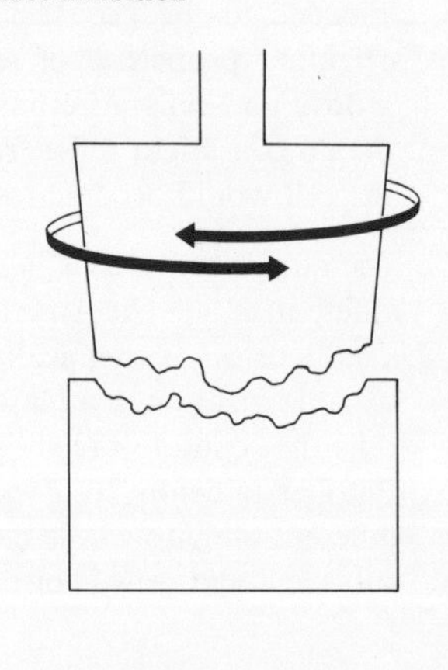

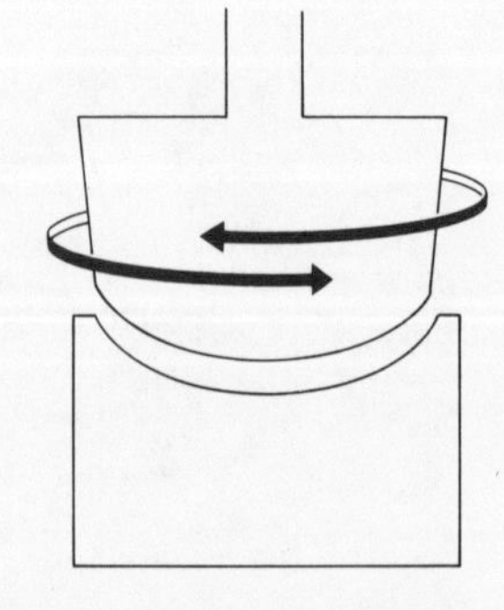

FIGURE 31 The principle of the inherently precise machine used as a metaphor by Elsdale to explain fibroblast bundle formation. The grinding of two rough blanks of glass at random gives perfectly smooth spherical lens surfaces.

Summary of Section 4

1 Rather than being static 'jigsaw puzzle pieces', an animal's cells are capable of movement during development.

2 The movement of the cells may be amoeboid or 'gliding'. The latter kind of movement is seen in fibroblasts in tissue culture.

3 Under culture conditions, cells can align themselves perhaps because of characteristics of the cells themselves (as suggested by Elsdale), or by properties of the matrix they are growing on (Weiss). This alignment may be the trigger for future differentiation, as in muscle tube formation.

Objectives and SAQs for Section 4

Now that you have completed this Section, you should be able to:

* describe how fibroblasts and other specified cells move and are oriented *in vitro*.

* list the properties of single cells that may be important in morphogenesis.

To test your understanding of this Section, try the following SAQs.

SAQ 8 (*Objective 9*) Which of the following statements are true and which false?

(a) Amoeboid movement does not involve a change in cell shape.

(b) Fibroblast movement involves a locomotory organelle called the leading lamella.

(c) Contact guidance is an orientation of cells activated by special characteristics of a macromolecular cellular secretion.

(d) Contact inhibition is the inhibition of cell movement between two neighbouring fibroblasts in culture.

(e) The inherently precise machine is used as a metaphor to explain amoeboid movement.

SAQ 9 (*Objectives 9 and 10*) How do cells *in vitro* respond to contact with the substratum and neighbouring cells?

5 Morphogenesis in cell populations

In the previous Section, we looked at the behaviour of individual cells. How do cells cooperate and interact to form specific shapes? In this Section, we will deal with the phenomenon of cellular reaggregation, and discuss types of recognition signals that may exist between cells.

5.1 Cellular reaggregation

The following experimental method was first used by H. W. Wilson at the beginning of this century. By squeezing pieces of sponge tissue through fine silk, he prepared a suspension of disaggregated cells from the sponge *Microciana prolifera* and allowed them to settle at the bottom of a dish. Soon, active movement of the individual cells led to the formation of numerous small clumps or aggregates of sponge cells. After about a day, some of the aggregates appeared as small but fully differentiated sponges, with an internal organization identical to that of the normal adult. The mechanism underlying this reconstitution soon became the subject of some controversy. Did the cells *sort out* according to their original type, or did their differentiated state change, so that cells on the outside of a reaggregated clump, for example, would become cells normally associated with that position even if they were originally inner cells. It was difficult to resolve this problem because the histological identification of the different types of sponge cells is not easy and they change their state and function in normal, undisturbed life (Unit 1, Section 3.3.2).

Fifty years later, Holtfreter and Townes discovered that the cells of an amphibian embryo will dissociate when exposed to saline solution at pH 10 and will reaggregate at pH 8. They tested the reaggregation of different combinations of ectoderm, mesoderm and endoderm cell suspensions. Because these types of cells are readily identifiable, the movement of each type of cell could be followed within the aggregates. Initially, the cells were distributed haphazardly, but by their subsequent movement, each cell type eventually sorted out according to type, and *each cell type took up a characteristic position within the aggregate* (see Figure 32). The separation of the various cell types seemed to mimic the situation seen in the real embryo. For example, if ectodermal and mesodermal cells are mixed, after reaggregation, the mesodermal cells form an inner core around which is a boundary of ectodermal cells. How does this sorting out occur?

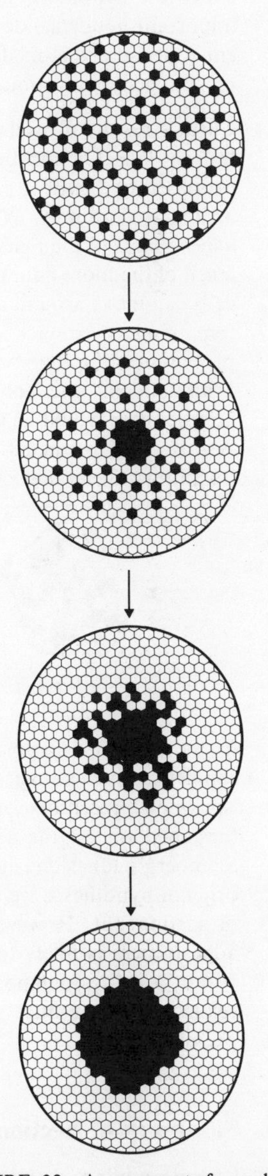

FIGURE 32 An aggregate formed from dissociated cells of two types arranged randomly; the cells sort out according to type giving the configuration of a sphere within a sphere. The same configuration can also be obtained when the two cell types are combined as tissue fragments.

5.2 The cellular basis of sorting out

Although sorting out, like the regeneration experiments mentioned earlier, is an *in vitro* phenomenon, it is important to establish the mechanisms at work. The underlying mechanism may well be significant in normal morphogenesis: the ability of cells to move into particular positions during development seems to be of importance, as we see later when we discuss morphogenesis in the embryo. Explanations of sorting out have taken three main forms. First, Holtfreter and Townes proposed a *chemotactic theory*, suggesting that the migration of cells or groups of cells occurs along a gradient of some metabolite. For example, if two cell types A and B are present and A cells produce some substance that helps A cells to aggregate, then it is possible that these cells will form a central mass and that B cells will be left to aggregate 'as best they can' around the central A cell core. Firm evidence against chemotactic theories is lacking, but there is no positive evidence either, so they are rather mistrusted.

chemotactic theory

A variation of this idea, and one that was popular for a time was the *timing hypothesis* of A. S. G. Curtis. Curtis worked with sponges, and noticed that different sponges took different lengths of time to reaggregate after dissociation. After mixing cells of two different sponge types, he observed that the faster reaggregating sponge species tended to form the central core of the reaggregant. This led to the hypothesis that it was the time at which cells reaggregated that was important in normal development—perhaps different genes switched on at different times and this made certain types of cell selectively 'sticky' possibly aiding morphogenesis in some way.

timing hypothesis

Curtis's results proved difficult to repeat, so the timing hypothesis has fallen by the wayside. The most elegant explanation for the sorting out phenomenon was proposed by Steinberg. Consider a mixed aggregate containing two types of cells arranged randomly. What happens if there is a quantitative *difference* in the *adhesiveness* of the two cell types with respect to cells of their own type? Imagine a cell of the more adhesive type A surrounded by less adhesive, B cells. Now if the cells can move around at random, then the probability of two A cells meeting and remaining in contact is higher, because of their greater adhesion, than is the probability of either two B cells or an A cell and a B cell. Steinberg described arithmetically the various possibilities for selective adhesion—for example, another case would be where the greatest adhesion is between cells of unlike type, A + B—and predicted the sorts of cell groupings that would occur for each possibility. Several examples are shown in Figure 33. Much work has been done

cell adhesiveness*

(a)

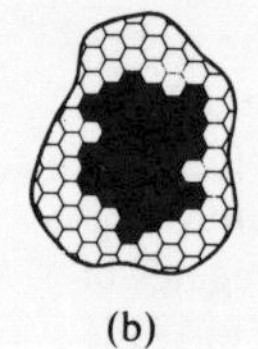

(b)

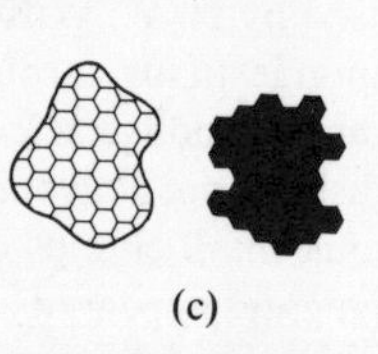

(c)

FIGURE 33 Relative adhesion between cells of two types. (a) The situation where maximum adhesion is between unlike cells. (b) The pattern formed if the black cells are the more attracted to one another. (c) No adhesion between unlike cells.

to try to find experimental evidence for Steinberg's hypothesis, ranging from sophisticated histological studies (to look for possible candidates for the role of 'stickiness' receptors on the cell surface) to complex computer simulations (designed to test the behaviour of abstract 'computer cells' programmed with Steinberg's adhesion rules'). Now fifteen years after the formulation of Steinberg's original hypothesis, we are still no nearer to a full explanation of the mechanism of sorting out. However, two general conclusions seem warranted. First, cell adhesiveness is likely to play a major role in establishing the position of a cell within a cell population, and second, morphogenetic movements may well be related to quantitative differences of adhesiveness of various cell types within a population.

Summary of Section 5

1 Both adult and embryonic cells can be dissociated and persuaded to clump together into a mixed aggregate where different cell types are randomly distributed. Subsequently, cells sort out as they move within the aggregate to become surrounded by cells of their own type. Each reconstituted tissue eventually adopts a characteristic position within the aggregate, even when the two cell types are combined as tissue fragments.

2 The behaviour of ectodermal, mesodermal and endodermal cells during aggregation bears some resemblance to their normal arrangement in the embryo. Sorting out may, therefore, have considerable morphogenetic significance.

3 Sorting out appears not to occur by chemotaxis, although this cannot be ruled out altogether. Steinberg's hypothesis proposes that sorting out in an aggregate can be accounted for by quantitative differences in cell adhesiveness. Adhesions between unlike cells are exchanged for adhesions between like cells, and the more adhesive cells move towards the centre.

4 If Steinberg's ideas are correct, it would mean that ectoderm, mesoderm and endoderm cells have different degrees of adhesiveness and these differences could well be of great importance in morphogenesis.

Objective and SAQs for Section 5

Now that you have completed this Section, you should be able to:

★ describe critically the various hypothesis proposed to account for sorting out.

To test your understanding of this Section, try the following SAQs.

SAQ 10 (*Objective 11*) (a) If a mixture of ectoderm and mesoderm cells are allowed to undergo reaggregation, which would you expect to take up the core position? (b) How does the 'timing hypothesis' explain sorting out?

SAQ 11 (*Objective 11*) Three types of tissue, A, B and C, are isolated and allowed to reaggregate in pairs. If B cells form a ball inside an outer covering of C cells, and C cells take up the core position when mixed with A cells, what would you expect to be the result of mixing A and B cells?

6 From pattern specification and cell properties to morphogenesis

So far, we have looked at two main things. First, we have tried to see how spatial arrangements are set up during development, or how patterns are specified. We have seen that the basic mechanism of pattern formation (if one exists) is not known, although theoretical ideas like the French flag hypothesis, and the idea of gradients in general, have their importance strengthened by their applicability to the analysis of experimental systems like *Anabaena* and *Rhodnius*.

Given that positional information is available to the various cells in a developmental field, it is then necessary to explain how the cells use this information to cooperate to build the adult organism by morphogenesis. Before considering the complex structures themselves, which are the subject of the following Sections, we looked at the properties of both individual cells and simple cell populations: such properties as cell movement and cell adhesion were described and their importance to specific developmental systems are described later.

It is important to emphasize the very inadequate foundation on which our knowledge of the pattern formation process lies. It would obviously be useful if we could directly describe what happens in the terminology of biochemistry or molecular biology, but this is simply not possible at present. Although we tend to think in terms of gradients of some biochemical substance as pattern specifiers, we have no hard evidence of the existence of any chemical morphogen: in the past few years, a group working on regeneration in *Hydra* have reported isolating a low relative molecular mass substance, which they claim is the *Hydra* morphogen, but further work will be necessary before this is proved.

Having looked at the 'building blocks' of morphogenesis, the cells, we can now try to unravel the process of multicellular morphogenesis itself. What cellular properties are important in shaping the developing embryo? What experiments have been done to study such problems?

First, we examine two developing systems that tell us something about the importance of particular cell properties during development. The first, gastrulation in the sea-urchin, gives insight into the role of cell adhesion and cell shape in morphogenesis, while the same process in an amphibian seems more complex.

We then turn to the control signals which switch on and direct morphogenetic events. This is an area that has fascinated biologists for most of the present century. What evidence is there that such signals exist. What form do they take? This leads us to look at what is perhaps the most exciting area of present day developmental biology: the role of the genes in development. This Section takes us back to *Drosophila*, the developmental geneticist's organism *par excellence*.

7 Gastrulation

You may recall from Unit 11 that the formation of the gut during embryogenesis is called *gastrulation*, and consists of an invagination process.

gastrulation

7.1 Sea-urchin gastrulation

In the sea-urchin, the fertilized egg gives rise, by a series of cell divisions, to a hollow ball of cells called the blastula, and it is the morphogenesis of the blastula into the *gastrula* that we are concerned with here. Figure 34 shows the general features of the process involved. *Study this figure now, and then attempt the following questions.*

FIGURE 34 Gastrulation in the sea-urchin embryo. Read through the annotation in order, stages 1–8.

① The blastula contains relatively few cells (1 000–2 000) arranged as a single layer around the blastocoel. At their outer surface, cells are attached to the **hyaline layer**.

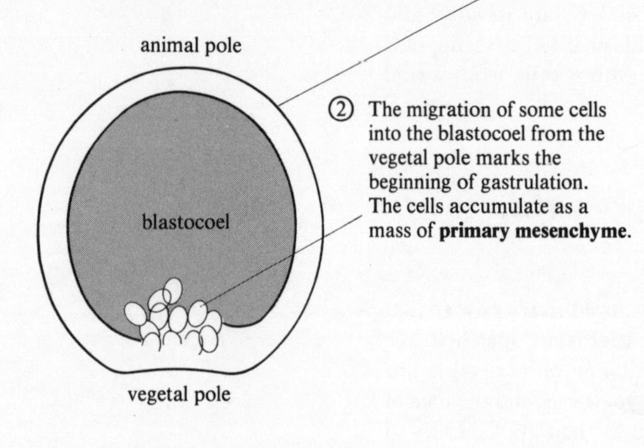

③ The remaining sheet of cells folds inward to form a hemispherical structure. This is accompanied by a 'rounding-off' of the inner surface of the vegetal cells.

④ The indentation formed is the archenteron .... with an exterior opening called the **blastopore**.

⑤ Primary invagination ceases after about 2 hours. Following a short delay, secondary invagination begins ....

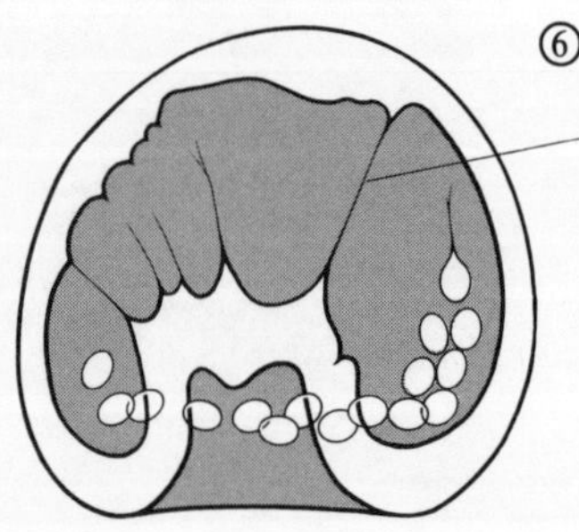

⑥ Cells at the archenteron tip extend fine processes termed **pseudopods** through the blastocoel fluid and anchor themselves mainly in the blastocoel wall.

⑦ The archenteron is pulled towards the animal pole and invagination is completed.

⑧ In later development the archenteron becomes the larval gut .... and the primary mesenchyme forms the basis of the larval skeleton

ITQ 1 What events does Figure 34 suggest are responsible for archenteron formation?

ITQ 2 What different cell types are shown in Figure 34?

ITQ answers and comments begin on p. 58.

In the previous Sections, we have looked at two properties of cells *in vitro*, shape and adhesiveness, and have hinted that these properties may be of importance during morphogenesis. From an analysis of echinoderm development, *Gustafson and Wolpert* proposed that the events of sea-urchin gastrulation could be accounted for by the two cellular properties, changes in cell adhesiveness and cell shape.

Gustafson & Wolpert's theory

First, let us concentrate on changes in adhesiveness. Gustafson and Wolpert's argument was as follows. Consider an idealized cell in contact with a flat base (Figure 35). If there is little adhesion between the cell and the underlying base,

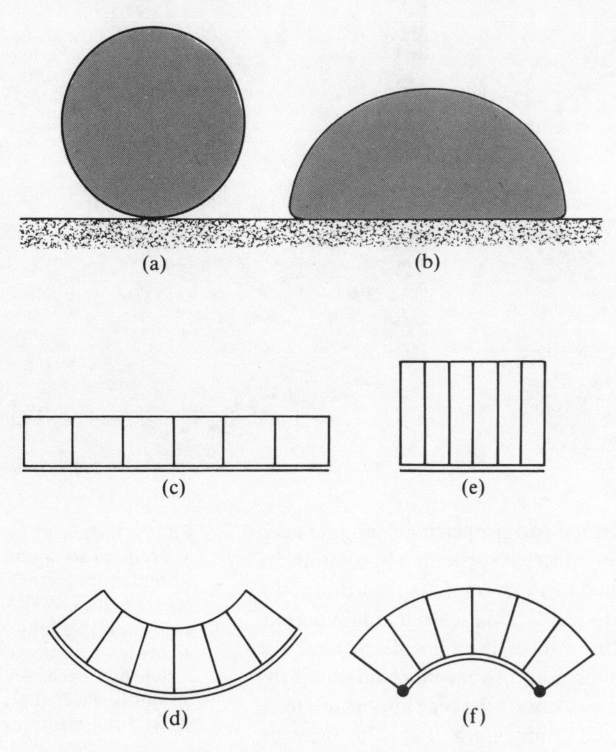

FIGURE 35 (a) and (b) Change in cell shape that might accompany a variation in the adhesiveness of the cell. (c)–(f) The effect of changes in cell adhesiveness on the form of a sheet of cells.

then the cell will not have much contact with the base and (at least in the case of our idealized cell!) will be rounded in shape (Figure 35a). If the amount of adhesion increases, then there will be more contact between the cell and the base, and the cell will become flatter by stretching (Figure 35b). Hence the extent of contact between a cell and its base (or perhaps between two cells) depends on a balance between the forces that tend to increase mutual contact (e.g. increased adhesiveness) and the forces that resist deformation.

If we take the argument one step up from single cells to sheets of cells, can we produce specific tissue shapes by this mechanism? In the sea-urchin, the cells at the vegetal pole (where the invagination to form the embryonic gut starts) are attached to a membrane called the *hyaline layer* on their outer edge. If we consider that the adhesion between the cells is moderate, a diagrammatic representation of the cells and membrane might look like Figure 35c. Changes in cell adhesiveness lead to changes in shape. For example, if the cells become more adhesive to both themselves and to the membrane, then Figure 35d would show the resulting situation. If the cells became more adhesive only to themselves, then Figure 35e would be the result.

hyaline layer

Gustafson and Wolpert proposed that the cells at the vegetal pole before invagination resembled those in Figure 35e. They suggested that contact between adjacent cells is then reduced, but contact between the cells and the hyaline layer remains the same. If the ends of the cell sheet are fixed (by some characteristic of the wall of the invaginating blastula), the sheet cannot spread so it curves. This situation is shown in Figure 35f. The curve this time is *inwards*, forming the *archenteron*, a cavity that later forms the gut. Gustafson and Wolpert's model therefore suggests the 'primary invagination' could result from a change in the adhesiveness of certain cells at the vegetal pole.

archenteron*

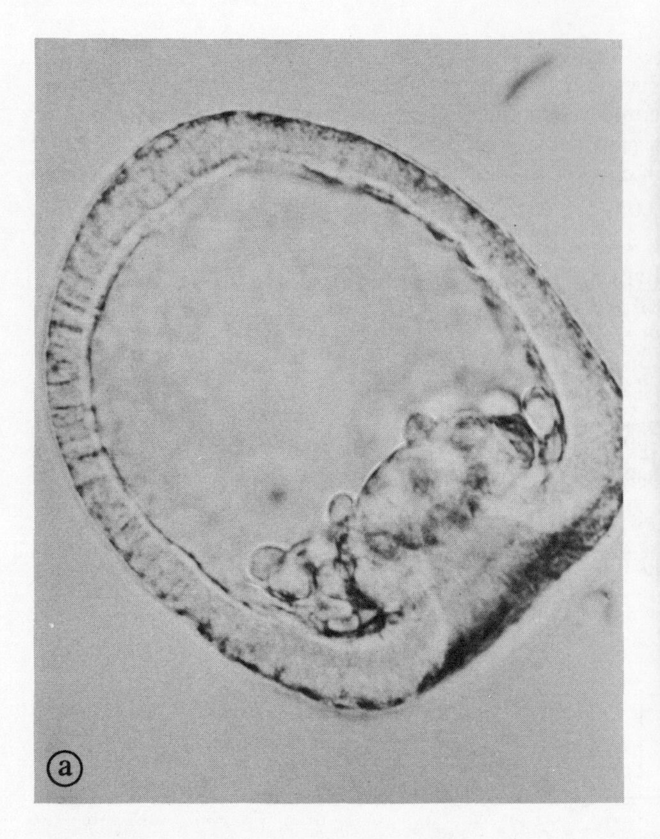

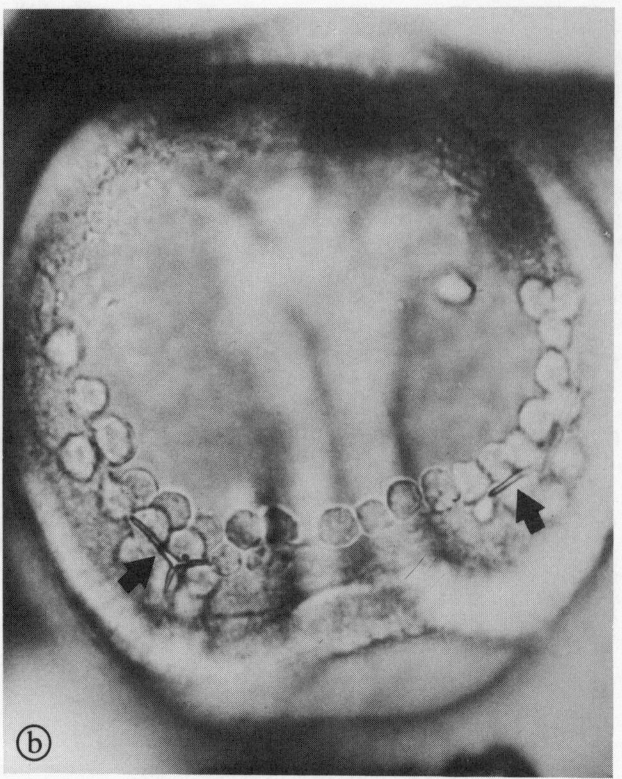

FIGURE 36 (a) An early sea-urchin gastrula showing primary invagination with primary mesoderm cells as a random cluster. (b) A late sea-urchin gastrula with a fully invaginated archenteron. A ring of primary mesenchyme cells is evident with two upwardly directed branches. Two skeletal rudiments are arrowed.

Turning now to consideration of the second of the two properties, changes in cell *shape*, we can look at what happens when the *primary invagination* has got under way. The second phase of gastrulation is marked by the formation of pseudopodia by the cells at the archenteron tip. These cells seem to have a stretching function, the pseudopodia anchoring themselves to the wall of the blastula around the animal pole (see Figure 36) and then contracting, and so mechanically help the elongation of the invaginating gut. This process seems to be very important, for if the blastocoel is treated with sucrose solution, producing a change in pressure inside the cells, the pseudopodia break down and archenteron invagination stops.

secondary invagination*

With the accent on changes in cell adhesiveness during the primary invagination of the gut, and cell shape changes with pseudopodial formation during *secondary invagination*, Gustafson and Wolpert underlined the importance of simple cell properties in seemingly complex morphogenetic events.

7.2 Amphibian gastrulation

Can simple cell properties explain morphogenesis in vertebrates as well? Most of the material to be covered in this Section deals with amphibians, and we restrict ourselves to a consideration of amphibian gastrulation, which seems more complex than gastrulation in the sea-urchin.

How does vertebrate gastrulation differ from that of the sea-urchin? In the 1940s, Holtfreter put forward an explanation of the role of individual cells in amphibian gastrulation. The general features of this process are shown in Figure 37, which you should study in some detail.

ITQ 3 What three features are shared by sea-urchin and amphibian gastrulation? (Compare Figures 34 and 37.)

surface coat

Holtfreter stressed the morphogenetic importance of a continuous *surface coat*, which he claimed was established around the amphibian egg before fertilization. This coat was supposed to give close contact and communication between the peripheral cells of the embryo; for example, cell movements started at one place would be communicated to adjacent parts, and so the surface coat could therefore coordinate morphogenetic movements.

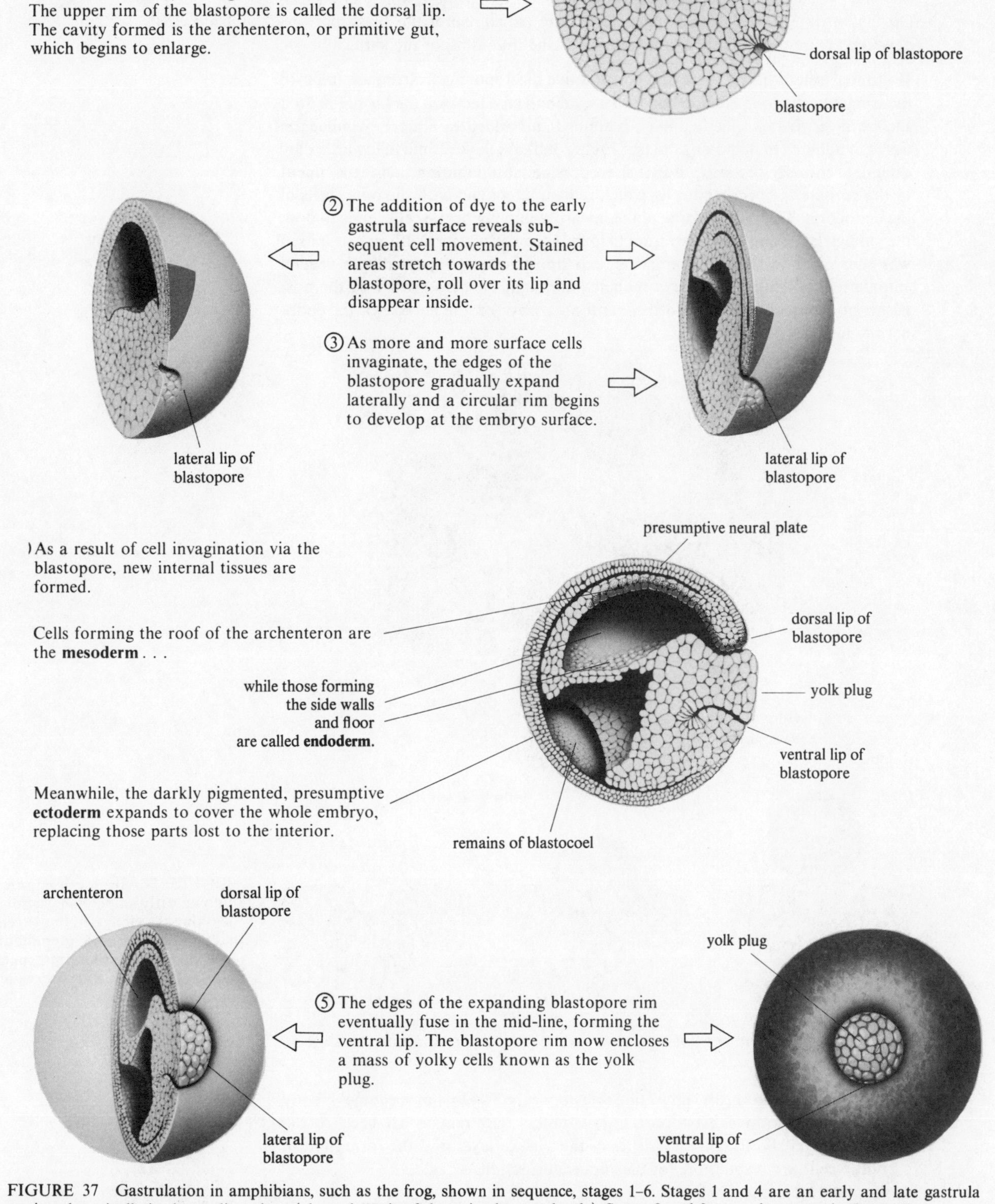

FIGURE 37 Gastrulation in amphibians, such as the frog, shown in sequence, stages 1–6. Stages 1 and 4 are an early and late gastrula sectioned vertically in the median plane (along the axis of the animal–vegetal pole). Stages 2 and 3 are embryos cut in the median plane, but viewed at an angle. Stage 5 shows a late gastrula viewed from the side bearing the yolk plug: to the left is an angled view of an embryo cut in the median plane; to the right a view of a whole embryo, showing the darkly pigmented ectoderm surrounding the non-pigmented yolk plug cells. As gastrulation ends, the blastopore rim contracts and finally covers the yolk plug altogether. So, all the material of the vegetal region eventually disappears into the interior of the embryo.

The surface layer has never been conclusively identified. The cells at the periphery of the blastula certainly seem to be closely associated with one another, and are united in a single cell sheet. Holtfreter proposed that the special properties of the peripheral cells are important in the formation of the *blastopore* (Figure 38a). Holtfreter observed through the microscope that as invagination begins, the cells connected to the invaginating gut become elongated, stretching into the blastocoel (in rather the same manner as the sea-urchin pseudopodia we saw in Section 7.1). Do the *bottle cells*, as these elongated cells are called, *cause* the blastopore to develop, or are they simply a consequence of the unfolding of the surface?

blastopore*

bottle cells*

Holtfreter tested this by removing presumptive blastopore cells from an embryo: the isolated cells soon rounded up into a solid ball covered by a darkly pigmented surface layer. If this spherical mass is stained and added to a piece of *endoderm* tissue, it adheres to it and sinks in (see Figure 38b). As the cells move inwards they extend, eventually becoming bottle shaped, while still retaining their attachment to the peripheral blastopore cells, which remain in contact with the outer cells of the endoderm. The pulling force of the inwardly moving bottle cells tends to drag the surface layer into the substratum to form a groove, which Holtfreter supposed was equivalent to the blastopore. The experiment can be repeated with vegetal endodermal gastrula cells as the implant—although the cells grow into the host tissue, no elongated cells are formed, and no groove on the surface of the endodermal host tissue is seen.

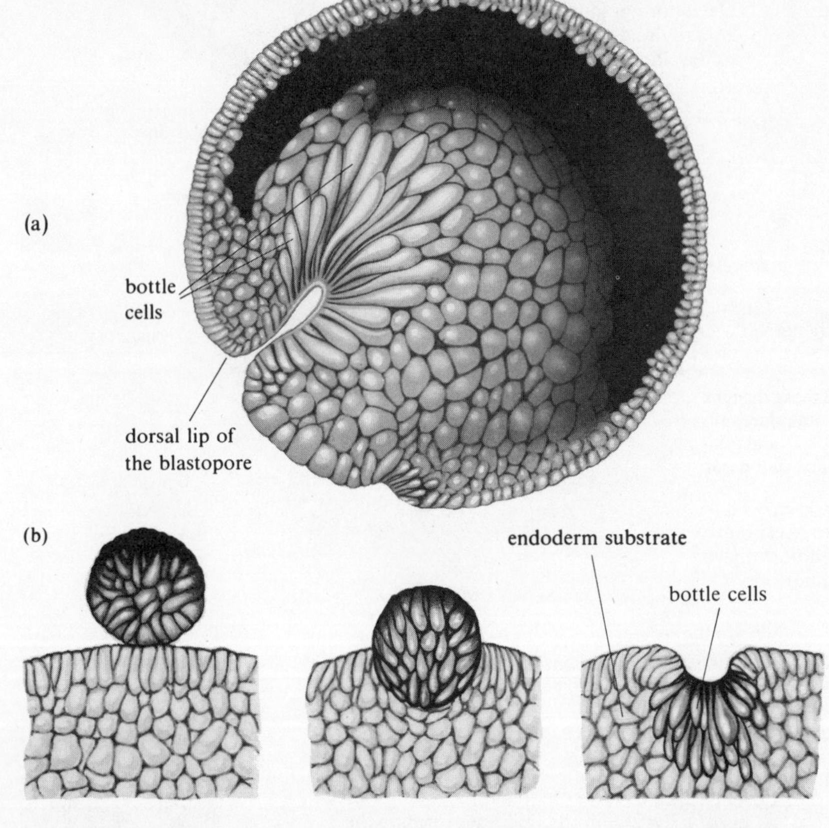

FIGURE 38 (a) Idealized side view into an early amphibian gastrula, showing the elongated bottle cells in section. (b) A graft of presumptive blastopore cells sinking into an endoderm substrate to form a groove, as revealed in cross-section.

□ From these experiments, what factors would you suppose initiate blastopore development?

■ It seems that inwardly migrating blastopore cells do not become greatly stretched or form a blastopore groove unless they remain attached to the surface layer. So the characteristics of the surface layer and the formation of bottle cells are responsible for blastopore formation.

Unfortunately, the picture is not complete—we have little idea about the mechanism responsible for the inward movement of the bottle cells. Possibly it is related to an increase in their adhesiveness, which according to Steinberg's theory (Section 5.2) might result in an inward movement.

□ What structure might integrate the pulling activity of the bottle cells?

■ The surface layer. Note that the same structure might also communicate the pull to other more distant cells.

From Figure 37, notice that following the formation of the blastopore by bottle cells, many of the surface cells begin to move towards the expanding blastopore, then roll over its lip and disappear inside. What cellular forces are responsible for this type of movement?

First, notice that the darkly pigmented ectodermal cells spread during gastrulation (Figure 37).

□ If the entire ectoderm is removed from an embryo, cells still move towards the blastopore and invaginate more or less as normal. What does this result suggest about the mechanism of invagination?

■ It suggests that invagination via the blastopore does not result from a push exerted by the spreading ectoderm.

We know of at least two factors that do seem to be involved.

1 If certain parts of the surface of the gastrula are removed before their invagination and cultivated in saline media, the individual cells stretch. This inherent ability of cells to stretch in the appropriate direction may contribute to invagination *in vivo*. (Compare this with invagination in the sea-urchin.)

2 During gastrulation, the mesodermal cells of the archenteron roof spread on the inner surface of the ectoderm (immediately beneath the presumptive *neural plate*—see Figure 37). This spreading appears to be due to a difference in the adhesive properties of the mesodermal cells and the inner ectoderm surface. (Again, compare with sea-urchin invagination.)

neural plate*

□ The adhesive properties of the two cell types enables a spreading of mesoderm on the inner face of the ectoderm. Which is the less adhesive?

■ The mesoderm is less adhesive than the inner ectoderm surface, and so it would tend to spread.

Invagination, therefore, appears to be linked with both a spreading of the mesodermal cells of the archenteron roof and an inherent tendency of some surface cells to stretch. Again, coordination of these processes might be achieved by the surface layer.

Finally, we should briefly mention the spreading of the surface ectoderm at gastrulation. Although the ectodermal sheet spreads as a unit during gastrulation, the individual cells also have a tendency to spread. This is apparent if surface ectoderm cells after their dissociation are cultured on glass. At first, the individual cells adhere to the glass and soon afterwards begin to spread. It seems that the expansion of the ectoderm depends on the capacity of the individual cells to spread. Quite how cell spreading is achieved and how it is coordinated remains a mystery.

ITQ 4 Now try to summarize Holtfreter's work by identifying:

(a) The factors involved in the formation of the blastopore.

(b) One example of the importance of the adhesive properties of any cell layer.

(c) Two factors contributing to the later stages of invagination.

(d) Two primary functions of the surface layer.

(e) Two tissues that spread during gastrulation.

Summary of Section 7

1 Gastrulation in both vertebrates (e.g. amphibians) and invertebrates (e.g. sea-urchins) seems to involve simple cell properties like adhesion and shape.

2 The main features of sea-urchin gastrulation are:

(a) Migration of primary mesenchyme cells to the blastocoel from the vegetal pole.

(b) Primary invagination may involve a change in adhesiveness of the adjacent cell membranes at the vegetal pole.

(c) Pseudopodia of cells at the archenteron tip attach to the upper part of the blastocoel wall and then contract.

3 Invagination in the amphibian occurs by inward migration of the blastopore cells (bottle cells). Possibly the properties of the surface coat are also important in gastrulation, partly because explanted bottle cells sink into an endodermal substrate but do not seem to be able to anchor to the outside of this coatless cell layer.

Objectives and SAQs for Section 7

Now that you have completed this Section, you should be able to:

* briefly summarize gastrulation in the amphibian and sea-urchin embryos, and explain how cell properties such as adhesiveness and movement may be important in these processes.
* list the differences between gastrulation in simple and complex animals by comparing the sea-urchin and the amphibian.

To test your understanding of this Section, try the following SAQs.

SAQ 12 (*Objective 12*) Which of the following experimental observations (i)–(v) fit with Gustafson and Wolpert's theory about the mechanism for *primary* invagination in the sea-urchin?

(i) Pseudopodia of archenteron cells are attached to the animal pole and shorten as the archenteron elongates.

(ii) Some of the vegetal cells detach, migrate into the blastocoel, and accumulate as a mass of primary mesenchyme.

(iii) Changes of adhesiveness occur between cells at the vegetal pole.

(iv) The sheet of cells at the vegetal pole bends inwards to form a more or less hemispherical structure, the cavity of which is the archenteron.

(v) Suppression of pseudopodial activity prevents invagination.

SAQ 13 (*Objective 12*) Answer the following questions about amphibian gastrulation briefly.

(a) What factors initiate blastopore development?

(b) Does the blastocoel enlarge or contract as gastrulation proceeds?

(c) Which type of cells in the gastrula are darkly pigmented?

(d) What is the name of the primitive gut formed during gastrulation?

(e) What experiment shows the special invasive properties of the bottle cells?

SAQ 14 (*Objective 12*) Figure 39a–f shows the proposed effect of changes in cell contact in a cell sheet supported by an outer membrane. Which of the descriptions (i)–(vi) refer to each of the Figures (a)–(f)?

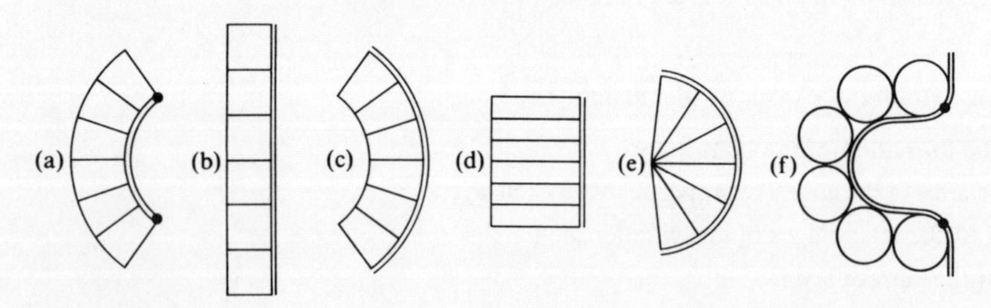

FIGURE 39

(i) Moderate contact between the cells and between the cells and the supporting membrane.

(ii) The effect of a further increase in contact between the cells, compared to that in (c), if the cells do not lose their contact with the supporting membrane.

(iii) The effect of the cells in (d) reducing their contact if the ends of the sheet are fixed.

(iv) The result of increased contact between the cells and reduced contact with the supporting membrane.

(v) The effect of rounding up the cells in (b) if the ends of the sheet are fixed.

(vi) The effect of increased contact between the cells if the cells do not alter their contact with the supporting membrane.

SAQ 15 (*Objective 13*) What are the main factors that make vertebrate gastrulation more complex than invertebrate gastrulation?

8 Embryonic induction

Blastula and gastrula formation are only the first steps in the chain of morphogenetic events that finally produce the adult organism. We have already looked at the sorts of mechanical properties that cells have to bring about these changes, and have seen how the accuracy with which we can describe morphogenetic events is correlated with the simplicity of the system being studied. Do we have any information about later stages of embryogenesis? Can we recognize any of the 'control signals' that initiate developmental events.

To some extent, this Section deals with a series of failures: a set of experiments that set out to explain developmental control mechanisms in terms of biochemical signals. The experiments are important for several reasons, and they show the *general* nature of developmental phenomena: tissue from one organism can often provide 'signals' for a tissue from another organism to develop normally. They also demonstrate that signals of some kind pass from tissue to tissue during development.

embryonic induction*
neurula*

In this Section, we discuss the phenomenon of *embryonic induction*, already introduced in Unit 11—in particular, development up to the stage of the *neurula*, the embryological term given to the developmental stage of the vertebrate embryo following the gastrula. At this time, nerve tissue first appears as a flattened, dorsally located plate: subsequently the edges of the plate fuse to form a hollow neural tube.

If a small piece of *presumptive* epidermis (a piece of embryonic tissue that would normally go on to form a particular tissue in the adult is called presumptive) from an early newt gastrula is transplanted into the area of another embryo that is destined to form neural plate, the transplanted piece develops in conformity with its new surroundings, first as neural plate tissue and then as part of the neural tube. In the same way, presumptive neural plate from an early gastrula differentiates as skin epidermis following its transplantation into presumptive skin epidermis of a similar gastrula (Figure 40). Evidently the neural plate tissue could not have been determined at the time of transplantation: its fate was eventually set by some factor related to its new position.

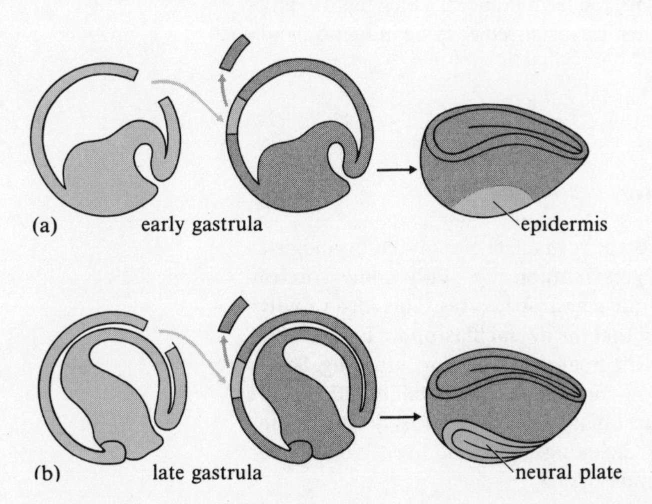

FIGURE 40 (a) Presumptive neural plate removed from an early gastrula and transplanted into a host gastrula at the same stage develops as belly epidermis. (b) A corresponding piece removed from a late gastrula differentiates as neural plate, according to its original fate. Host tissue is grey, donor tissue is pink.

If similar experiments are carried out on late gastrulae, the results are strikingly different. A transplanted piece of presumptive neural plate develops as neural tissue irrespective of its new position; transplanted presumptive epidermis develops into epidermis despite its new location within the presumptive neural plate (Figure 40b).

By the end of gastrulation, the neural plate has lost its former ability to develop into epidermis. How is the neural tissue determined? The transplantation studies described above show that each cell is not endowed with a set role from the moment of its creation. Some evocative signal from the surrounding cells is required before differentiation can proceed normally—the determination of the neural plate thus provides an example of embryonic induction.

It is now nearly 60 years since the first experiments to further characterize the embryonic induction process. Mangold and Spemann showed that the roof of the archenteron influences the determination of the neural plate. Mangold did a fascinating transplant experiment with two different species of newt, one colourless and the other heavily pigmented. When the *dorsal lip of the blastopore* (Figure 41) was transplanted from the early gastrula of one species into the

dorsal lip of the blastopore*

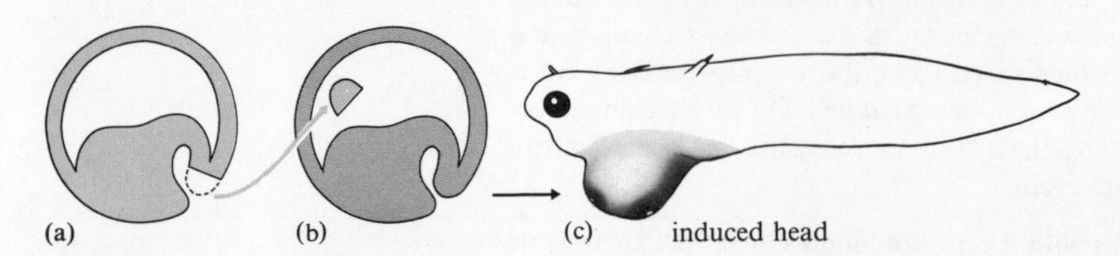

FIGURE 41 Transplantation of blastopore material from lightly pigmented gastrula (a) into ectoderm of darkly pigmented gastrula (b) gives rise to (c) twin embryo development. The pigmentation pattern of the induced second embryo is that of the host and not of the graft, suggesting that the implant's role is *organization*.

ectoderm of the gastrula of the differently coloured species, a *second embryo* started to develop on the implant site, containing a neural tube as well as an embryonic *notochord* and blocks of tissue destined to become muscle (Figure 41). Because the second embryo was formed from tissue pigmented like the host and not the implant, the grafted cells therefore must have given some signal for the host cells to change their fate. How was this possible? The problem was made worse when it was realized that the transplant could not only influence presumptive neural plate, but also endoderm and mesoderm. Spemann subsequently called the dorsal lip of the blastopore the *organizer*, this term being meant to imply an inherent ability to organize the whole process of development. The problem of the organizer, and of embryonic induction in general, dominated experimental embryology for almost 30 years following the organizer's discovery in 1924. Most embryologists believed that if the substances responsible for the inductive powers of the organizer could be isolated, the key process of development could be identified. The fallacy in this argument is that even if an 'organizer substance' could be isolated (which has not been done to date), it would do very little to explain how the complex processes of development produce shapes and patterns.

notochord*

organizer*

We use the term *inductor* for a substance or tissue that influences a second tissue during development. Do not confuse this with the term inducer, which has already been used in a different way in Unit 11, when we discussed enzyme induction and repression.

inductor*

8.1 The chemical identity of the inductor

In the naivity of their initial rush to find the 'organizer', many embryologists believed that it must be a chemical of great complexity—only some sort of 'master-chemical' could induce such elaborate neural structures. This idea rapidly turned out to be wrong. First, it was found that the dorsal blastopore lip was able to induce neural plate even after quite drastic treatment: freezing, crushing, heating, or by treatment with various kinds of solvents such as alcohol, ether and chloroform. Next, attempts to induce neural plate with substances derived from other tissues were found to be successful, including vertebrate liver, kidney and muscle derivatives, besides some invertebrate tissues.

Some experiments seemed to indicate that a chemical is involved. Niu and Twitty incubated archenteron roof or dorsal lip tissue in special saline solution. After ten days incubation, the inducer tissue was removed and a very small piece of presumptive neural plate was placed in the remaining *conditioned medium* (i.e. medium that has had some kind of 'factor' provided by an incubation such as that described). Following 24 hours exposure, most of the tissue fragments subsequently differentiated into neural plate. Control presumptive neural plate tissue incubated in *unconditioned* medium remained as a sheet of undifferentiated cells. This shows that in neural plate induction at least, the inductive stimulus appears to be chemical. Without this kind of experiment, it would be difficult to rule out the possibility that the stimulus could be provided by some kind of physical interaction between neighbouring cells.

conditioned medium

unconditioned medium

The situation was shown by Niu to be even more complicated than a simple search for a single chemical that could do one specific job. If the archenteron roof of embryos that were 7–10 days old was used as the inductor, the receptive neural plate cells developed mainly into nerve tissue. If the archenteron roof was derived from embryos that were 12–15 days old, muscle cells were a major product. These experiments raise the problem of there possibly being a range of inductor molecules produced at different times in development and responsible for different developmental activities.

Many embryologists thought that finding out the chemical nature of the sorts of substances that bring about induction would give clues to the process itself. Spemann discovered that inductive ability is not limited to the dorsal lip of the blastopore and the archenteron roof. Neural tissue, following its own induction, can itself induce neural plate differentiation if transplanted beneath presumptive neural plate. This ability is retained during development: for example, spinal cord of amphibian larvae and adults can also induce. Non-neural embryonic tissue (e.g. notochord) has the same effect. Vertebrate liver, kidney and muscle cells are also effective. With the newt as a host, various adult tissues from a variety of animal species act as inductors—*Hydra*, insects, fishes, reptiles, birds and mammals.

normal and abnormal inductors*

Niu tried to identify the active agent in the conditioned medium. Elaborate chemical analysis proved difficult because of the minute quantities of inductor released. Nevertheless, he succeeded in extracting a ribonucleoprotein (a complex of ribonucleic acid and protein) with strong inductive activity, and concluded that the inductive properties of the substance lay in the RNA portion of the nucleoprotein. Niu's claim has been disputed by other workers.

8.2 A further complicating factor: a whole range of inductors

Work with abnormal inductors has shown that different tissue preparations have varying inductive abilities. For example, liver and kidney tissues exclusively induce neural structures: in other words, they are *neuralizing agents*. Conversely, alcohol-treated bone marrow of the guinea-pig induces mesodermal parts of the trunk and tail: it is said to be a *mesodermalizing agent*. The situation is further complicated by the fact that certain tissue extracts induce specific neural structures. For example, alcohol-treated liver promotes differentiation of anterior neural structures (fore-brain, eye and nose rudiments), while the kidney of the adder (!) specifically induces hind-brain and ear rudiments. The alcohol-treated kidney of the guinea-pig induces spinal cord, notochord and muscles.

neuralizing agent*

mesodermalizing agent*

Saxen and Toivonen studied this problem of the multiplicity of inductive effects. They implanted two tissue pellets simultaneously into the blastocoel of a young newt gastrula. One pellet, prepared from guinea-pig liver is on its own only able to produce anterior neural structures—fore-brain, eye and nose rudiments. The other pellet, of bone marrow, gives mesodermal induction—notochord and muscles—if present alone. Implanted together, the complete range of neural structures were formed—fore-brain, mid-brain and hind-brain together with the ear rudiments and spinal cord. These results were explained by there being three different areas of competent ectoderm. Two respond to neuralizing and mesodermalizing influences, while the third requires an interaction of the two signals.

Saxon and Toivonen's interpretation of induction

There seems to be a chemical difference between the mesodermalizing and neuralizing agents. Neuralizing factor is soluble in organic solvents and is relatively stable during temperature changes. Mesodermalizing factor is insoluble in organic solvents and readily breaks down on heating. Exhaustive chemical analysis has provided some clues about their identities; the mesodermalizing agent appears to be a protein, while in contrast there is some evidence (not supported by its thermostability) that the neuralizing agent isolated from guinea-pig liver might be a ribonucleoprotein.

In Section 2, we discussed how chemical gradients might set up patterns and hence direct morphogenesis. Do the results with the abnormal inductors described here give evidence for a gradient mechanism in the determination of neural structures? Saxen and Toivonen concluded that the full range of neural structures could only be produced if the type of structure formed depended on the interaction of each cell with both inductor substances: they proposed that the neuralizing and mesodermalizing agents are distributed unequally within the archenteron roof of the normal embryo in the form of opposing gradients

(Figure 42). The neuralizing agent was hypothesized as decreasing in concentration laterally and ventrally, while the mesodermalizing agent is absent in the most anterior part but its concentration gradually increases posteriorly.

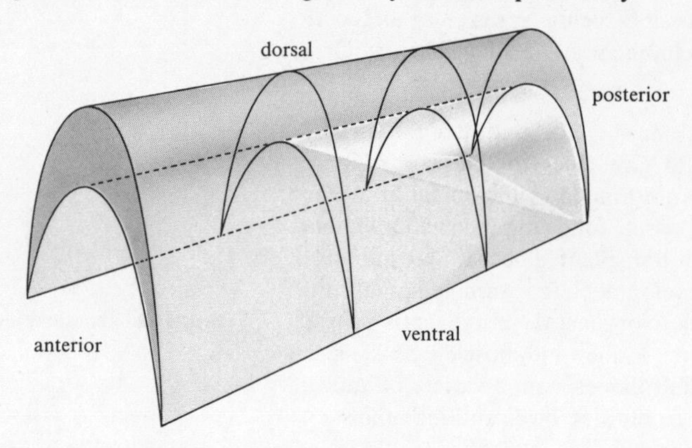

FIGURE 42 A schematic representation of Saxen and Toivonen's two-gradient hypothesis indicating the form of the gradients in the archenteron roof. Neuralizing factor is shown as a grey shade, mesodermalizing factor as pink.

Note that the central portion of the archenteron roof contains both neuralizing factor and small amounts of mesodermalizing factor. It therefore induces both neural (hind-brain) and mesodermal (notochord) tissue. In addition, there is supposed a dorso–ventral variation in neuralizing agent: the highest gradient level appears to induce neural plate, a lower level gives rise to sensory nerves, the adrenal medulla and pigment cells. These tissues are collectively called the *neural crest* and are ectodermal in origin. Parts of the ectoderm not stimulated by the neuralizing factor differentiate as skin epidermis.

neural crest*

We now consider briefly the philosophy of the approach used by these embryologists. Development in higher organisms is a complex process: just consider the vast range of cell types and organizational complexity of something even so humble as a newt neurula. Is it really possible to make meaningful statements about how morphogenesis works in terms of chemical inductors? Even if particular chemicals are identified as key components in, for example, neurulation, it is difficult to see how we would be better able to understand how the process works. A hierarchy of processes are involved, starting with control through genes in the nucleus and ending with the physical differentiation of particular cells. Because organisms are made up of large masses of cells, we have to face the fact that understanding the causal factors behind an event such as neural plate formation may involve control on different levels and may not be directly reducible to chemical equations.

8.3 An *in vitro* model for induction

In recent years, attention has been focused on an interesting *in vitro* experimental system developed by Grobstein in the late 1950s that promises to throw light on the mechanism of induction. This is the study of *ectoderm–mesoderm interactions*—the interrelation between *ectoderm* and *mesoderm* cells in development and how new morphogenesis occurs by their cooperation. Interactions of this kind are called *secondary inductions* to distinguish them from the primary inductions of early embryogenesis. Ectoderm–mesoderm interactions occur in a variety of situations. A tooth is formed by the interaction of the enamel ectoderm and the tooth mesoderm, the type of tooth formed being determined by the latter. Likewise, organs consisting of glandular *epithelia*, for example, pancreas, salivary tissue, mammary tissue and kidney, depend for their formation on the interaction between epithelium and mesoderm. As an example, take the pancreas, where even at a very early stage of development, pancreas mesoderm may be replaced with mesoderm from many places in the body—a normal pancreas still results.

What is the basis of this interaction? Morphogenesis in the epithelium seems to be triggered by some 'inductive' influence from the mesoderm. The experimental approach mentioned above for the pancreas shows a lack of mesodermal specificity. How can we find out more about the induction?

It has been known for many years that kidney tubules are formed as a result of the interaction of kidney mesoderm and epithelium. Grobstein separated the kidney mesoderm from the epithelial bud attached to it and placed it on a *Millipore filter*, a special filter with pore size of 0.45 μm. A piece of spinal cord epithelium was sealed to the opposite side with agar. After incubation for a number of days in a suitable medium, kidney tubules were formed in the mesoderm (see Figure 43).

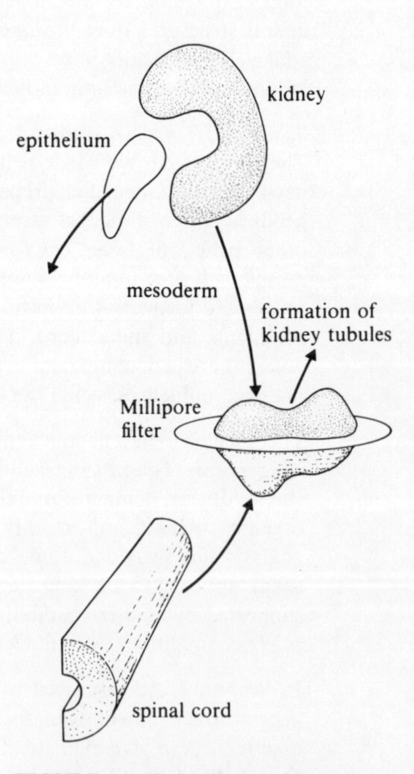

FIGURE 43 Epithelio–mesodermal interactions across a Millipore filter. Kidney mesoderm is separated from its adjoining epithelium and placed on a Millipore filter. Induction of kidney tubules occurs as a result of induction by apposed spinal cord tissue.

□ As there was a narrow-pored filter between mesoderm and epithelium, is the possibility of induction mediated by cell–cell contact ruled out?

■ No. Although the initial results of Grobstein's work were taken to imply that the inductive influence must occur by diffusion, electron microscopy showed that cell processes could enter the pores.

In fact, an inverse relation was found to exist between the pore size and the thickness of the filters across which the induction took place. Later, a minimum pore size of 0.15 μm was established for the inductive process to occur.

The transfilter method could also be used to measure the time needed for the inductive stimulus to be transmitted, because contact between the two sides could be broken at any time. From experiments of this kind, it was shown that the minimum induction period needed with one filter interpolated between the two tissue types was 12 hours. The diffusion rates of a number of substances were then tested with the filters. As all chemicals were found to diffuse across in less time than 12 hours, this was taken as extra evidence that cell–cell contacts must be formed within the pores.

We must therefore seriously consider direct cell–cell contact as a possible candidate for the inductive stimulus, at least for ectoderm–mesoderm interactions.

There is some information on what processes might be going on during ectoderm–mesoderm induction. It was noticed in electron micrographs of filters used for transfilter induction that collagen molecules (collagen is a protein, and a structural component of many biological tissues such as connective tissue and skin) are laid down in the pores during induction. Collagen is an integral component of cell basement membrane, and treatment with collagenase, an enzyme that degrades collagen, has been shown to affect the development of some ectoderm–mesoderm systems.

glycosaminoglycans

More evidence exists on the role of *glycosaminoglycans* (formerly called acid mucopolysaccharides), which are also present in the basement membrane. In the mouse embryonic salivary gland, freshly synthesized glycosaminoglycans accumulate on the surface of the epithelium, especially where there is epithelial branching in the mesoderm. If the epithelium of mouse embryonic salivary gland is treated with hyaluronidase (an enzyme that degrades glycosaminoglycans) and collagenase (to separate epithelial and mesodermal tissue), it loses its lobed appearance and becomes rounded. If it is cultured further with salivary mesoderm, it again becomes branched. *Now study Figure 44, in which the changes in epithelial morphology can easily be seen.* Biochemical analysis showed that the glycosaminoglycans were evenly distributed over the surface of the rounded explant, but on reacquisition of the branched pattern, these substances were again concentrated at the branching points. If this experiment is done with only collagenase in the separation procedure, the branching pattern is not lost.

hyaluronidase
collagenase

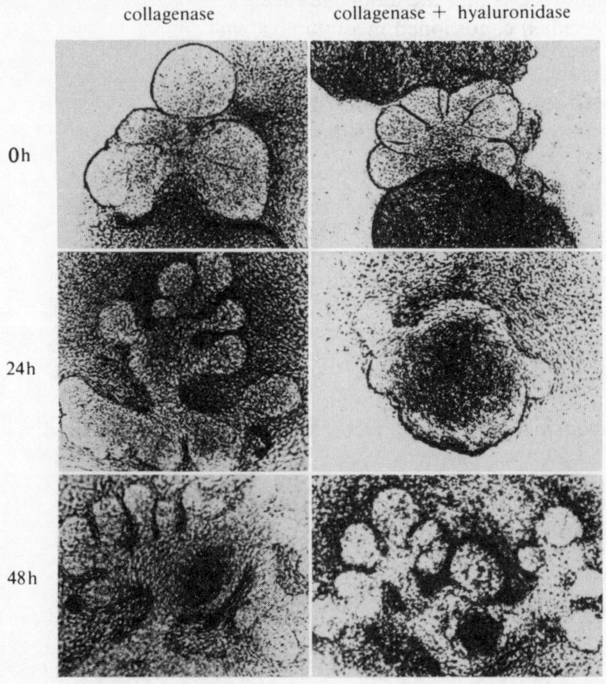

FIGURE 44 Mouse salivary gland epithelium shows normal development after collagenase treatment and subsequent growth in contact with salivary mesoderm. If treatment includes hyaluronidase, it initially rounds up, but later develops normally.

It seems that normal salivary gland morphogenesis depends on the glycosaminoglycans. The mode of action is not known. One speculation is that the contractility of microfilaments in the epithelial cells might require the presence of glycosaminoglycans. These microfilaments, it is suggested, may control cell shape and promote the formation of branches in the growing salivary gland.

8.4 Will the real inductor please stand up?

There are several candidates for embryonic inductors. The evidence from classical work on embryonic induction suggests a chemical substance or range of substances. Not only can a bewilderingly wide range of substances be used, from adder kidney to chloroform-treated blastopore lip, but also there is a time-dependent specificity, meaning that the 'substance(s)' must presumably change with time. (Remember Niu's conditioned medium experiments described in Section 8.1.) On the other hand, the ectoderm–mesoderm interaction story suggests a direct cell–cell interaction mechanism. Although chemical substances seem to be involved, they are heavy polysaccharide–protein complexes bound to the cell surface.

There is a third category of embryonic induction, mentioned earlier in this Unit, where some chemical or physical substrate is needed to promote *in vitro* muscle differentiation. In experiments on chick epidermis, where differentiation is normally controlled by the underlying mesoderm, the mesodermal effect can be simulated by embryo extract (i.e. homogenized embryos) provided that a physical substrate is available for growth. This substrate may be collagen or a Millipore filter.

Summary of Section 8

1 Embryonic induction is the process whereby cells of one type have their fate determined by cells of a second type. Some signal passes between the two types of cells.

2 Induction is commonly seen in early development, where blastopore tissue induces neural tissue from ectoderm.

3 The dorsal lip of the blastopore is called the organizer because of its ability to induce other tissues.

4 The search for the chemical basis of the inductor has occupied many embryologists for many years. Many types of embryo extract have been shown to have inductive ability, among them dorsal blastopore lip that has been frozen or crushed and substances derived from vertebrate liver or kidney.

5 The inductive effect may be transferred from embryonic tissue to culture medium after culture for a number of days. This suggests that the inductive effect is chemical in nature. The activated medium is called conditioned medium.

6 Different tissue preparations have different inductive abilities. These have been separated into neuralizing agents (e.g. liver extract, which induces only neural tissue) and mesodermalizing agents (e.g. guinea-pig bone marrow, which induces trunk and tail segments).

7 It has been suggested that gradients of neuralizing and mesodermalizing activity exist in the embryo and that cells respond to the relative levels of both.

8 Chemical analysis has suggested that neuralizing agent might be protein in nature, while mesodermalizing agent may be a ribonucleoprotein.

9 Ectoderm–mesoderm (or epithelium–mesoderm) interactions are suitable models for examining the mechanism of induction *in vitro*.

10 Typical interactions of this kind are responsible for forming teeth, kidney tubules, eye and pancreatic tissue.

11 Transfilter induction was first thought to demonstrate the biochemical nature of the inductor, but later work showed that cellular protrusions could negotiate the filter pores.

12 Glycosaminoglycans have been proposed as being important in salivary gland morphogenesis, an inductive process.

Objectives and SAQs for Section 8

Now that you have completed this Section, you should be able to:

★ summarize experimental approaches used in attempts to characterize the inductor and discuss the regional specificity of inductive effects.

★ summarize how ectoderm–mesoderm interactions provide a model for studying cell induction.

To test your understanding of this Section, try the following SAQs.

SAQ 16 (*Objective 14*) The following statements (a)–(e) are supposedly results obtained from experiments observing embryonic induction. Which are true and which false?

(a) When a graft of blastopore cells is made from an amphibian gastrula to a second gastrula of different pigmentation, the twin embryo formed has the pigmentation pattern of the grafted cells.

(b) Late gastrula neural plate develops according to its original fate when transplanted into a second late gastrula.

(c) Early gastrula neural plate develops according to its original fate when transplanted into a second early gastrula.

(d) Presumptive epidermis from an early amphibian gastrula can form neural tissue if transplanted into presumptive neural plate region of a second gastrula.

(e) The blastopore has a strong inductive effect on other regions of the embryo.

SAQ 17 (*Objective 14*) Identify the following material (a)–(e) as either abnormal inductor, normal inductor, or non-inductor.

(a) The dorsal lip of the blastopore of an early newt gastrula.

(b) A tissue extract of notochord from newt larva.

(c) The archenteron roof of a late gastrula.

(d) Alcohol-treated liver.

(e) Niu's unconditioned medium.

SAQ 18 (*Objective 14*) Slices of bone marrow were heated for varying times—25, 40, 60 and 150 seconds. Following each heat treatment, the inductive ability of the preparation was tested, and the resulting primary inductions were classified as A mesodermal, B hind-brain characteristics, and C anterior brain characteristics. Table 1 gives the induction obtained from each heat treatment.

In what way are these results related to Saxen and Toivonen's interpretation of primary induction?

TABLE 1 Percentage induction caused by heated bone marrow

Time of heat treatment	Type of induction		
	A	B	C
0 s	97	0	0
25 s	0	17	4
40 s	0	13	44
60 s	0	0	46
150 s	0	0	8

SAQ 9 (*Objective 15*) Are the following statements on ectoderm-mesoderm interactions *true* or *false*?

(a) Kidney tubules are induced by interaction between kidney mesoderm and spinal cord tissue.

(b) Transfilter induction needs to take place throughout the whole process of differentiation of the induced tissue to be effective.

(c) Cell surface molecules are thought to be important in transfilter induction.

(d) In ectoderm–mesoderm interactions, mesodermal tissue is influenced by epithelial tissue.

(e) Hyaluronidase inhibits the induction of salivary gland morphogenesis.

9 Genes and the control of development

We discussed evidence for the involvement of chromosomal material in differentiation in Units 12 and 13. If we accept that most of the differences arising between cells and tissues of a growing organism are due to the activity of heritable units of DNA, the *genes*, then we are faced with the problem of trying to explain the relationship between these genes and the shapes and patterns. Recent advances in developmental genetics have led to two main findings. First that individual genes can be responsible for controlling whole sequences of developmental events. Second, groups of cells derived from the same parent cell can differentiate during development to form different specific structures.

You read about genetic control mechanisms in Units 12 and 13, when we looked at the Jacob–Monod model for bacteria and its eukaryotic counterpart, the Britten–Davidson model. With bacteria, and even with some biochemical reactions in higher organisms (e.g. the hormone-induced synthesis of ovalbumin), it is possible to 'work out' what is happening at the biochemical level by deduction from experiments. Indeed, it was the intellectual *tour de force* involved in the elucidation of the mechanism of the *lactose* operon that led to the award of the Nobel prize to Jacob and Monod (see Units 12 and 13).

In cases like this, understanding of the genetic and biochemical mechanisms is not too distant. But what about the genetic control of morphogenesis? Little is known about the mechanism of shape and pattern formation, so how can we hope to have an understanding of their control? The answer is that knowing a little about the genetic basis of morphogenesis, or at least how tampering with genes may affect morphogenesis, can give us insight into the mechanisms themselves. Genetic techniques can also be used in their own right for studies on morphogenesis. They will be discussed in Section 9.3.

9.1 Mutants and development

If we take an abnormality, for example the *Drosophila* defect *vestigial wing*, we can establish whether or not this abnormality obeys the laws of heredity. These laws describe the behavior of independently assorting genes, and breeding experiments may be used to ascertain whether the vestigial character is associated with a particular gene on a particular chromosome. If a certain mutant gene is present, the fly has wings that are extremely reduced in size ('vestigial').

The single genetic difference between normal and vestigial-winged flies results in a huge difference in phenotype. Somehow the normal development of the wing must be affected by the genetic change. We can also deduce that the normal gene must play some part in the formation of the normal wing, because a change in the gene upsets wing development.

□ How could these observations be used to study how genes are related to developmental processes?

■ If we could describe how a given mutant allele affects normal development of a particular organ or group of organs, then this might throw light on the function of the normal allele.

mutants

With thc idea of using *mutants* to study the role of the genes in development in mind, Goldschmidt and Henke in Germany and C. H. Waddington in England tried to describe mutations at a physiological, histological and cellular level.

pleiotropy

A complication that can hamper our search for wild-type gene functions is the phenomenon of *pleiotropy*, where the initial genetic defect can interfere with several developmental pathways, so that the secondary effects produce complex mutant effects. One of the most studied examples of pleiotropy in humans is the Lesch–Nyhan syndrome produced by a recessive mutant gene. Affected persons are spastic, destructive and self-mutilating—but biochemical tests show that the genetic defect is the absence of a single enzyme!

Drosophila is an ideal organism for developmental studies, particularly of the growth of the imaginal discs. Many mutants have been found that interfere with disc development; among several thousand are examples that affect the overall shape of the wing (*dumpy*), affect wing size (*vestigial*), produce duplications of anterior wing structures in the posterior of the wing (*engrailed*), affect bristle structure (*javelin* or *singed*), affect bristle colour (*yellow* or *straw*), and affect wing vein pattern (*veinlet*). In addition, mutants can have more drastic effects. One group, called *homeotic mutants* transform whole sections of imaginal disc structures into those normally formed by other discs; for example, the mutant *ophthalmoptera* has wing tissue growing out of the compound eye, and *proboscipedia* has a leg growing out of the mouth parts. Two examples of mutant phenotypes in *Drosophila* are shown in Figure 45. The mouse mutant *nude* illustrates the use

homeotic mutants*

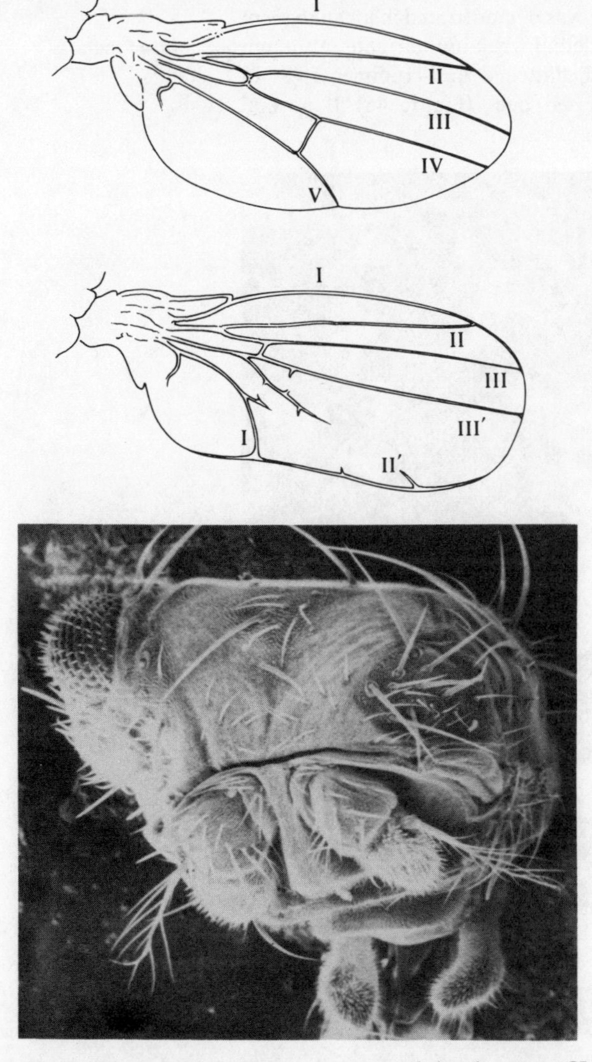

FIGURE 45 Examples of morphogenetic mutants of *Drosophila.* (a) Normal (top) and *engrailed* (below) wings. Wing veins are numbered with roman numerals to show the so-called duplication. (b) The head of an *eyeless* mutant. The eye on the left is drastically reduced in size. The eye on the right is missing and replaced by a patch of bristles.

of mutants in unravelling the link between genes and development. Homozygous *nude* mice are hairless, stunted in growth and infertile. Some years ago, it was shown that all of these effects result from the inability of the thymus gland to form early in development. In normal mice, the thymus is populated with cells (T cells) that subsequently migrate to other sites (e.g. the spleen and lymph nodes) and form lymphocytes. These cells produce an immunological response against foreign antigens, so the *nude* mouse with no thymus or T cells is incapable of rejecting grafts of foreign tissues.

In this example, the normal wild-type gene seems to be important in the morphogenesis of the thymus gland from the epithelium of the pharynx. By using standard biochemical and histological procedures, it is possible to track down what happens more precisely.

In this Section, we have looked at the possibility of using mutants to look at the workings of genes during development. This sort of study has traditionally been carried out with a variety of different techniques: histology, breeding experiments and surgical manipulations. But there are really two levels of genetic analysis—study of the phenotype of the whole organism, and study of the phenotype of groups of cells. We will next turn to the behaviour of cell populations during development and look at the evidence that individual cells may themselves take part in morphogenesis by acquiring differences in their genetic expression.

9.2 Mutations occur in cells

We have discussed the difficulty of unravelling primary and pleiotropic gene effects. Are there simpler examples of how genes control development? The answer is yes, because in some cases it is possible to directly analyse the behaviour of mutant cells in culture. By using this technique, the characteristics of cells as individual entities can be observed, and mutant and normal cell behaviour compared. It may then be possible to draw inferences about the role of these differences in bringing about normal or abnormal morphogenesis.

talpid mutants

One of the clearest examples of this technique was demonstrated in an analysis of the mutant *talpid* of the chick. This mutant is lethal when homozygous, although the embryos can be seen to develop broad, flattened limb rudiments (as in moles—*Talpa*) instead of the normal elongated ones (Figure 46). If normal

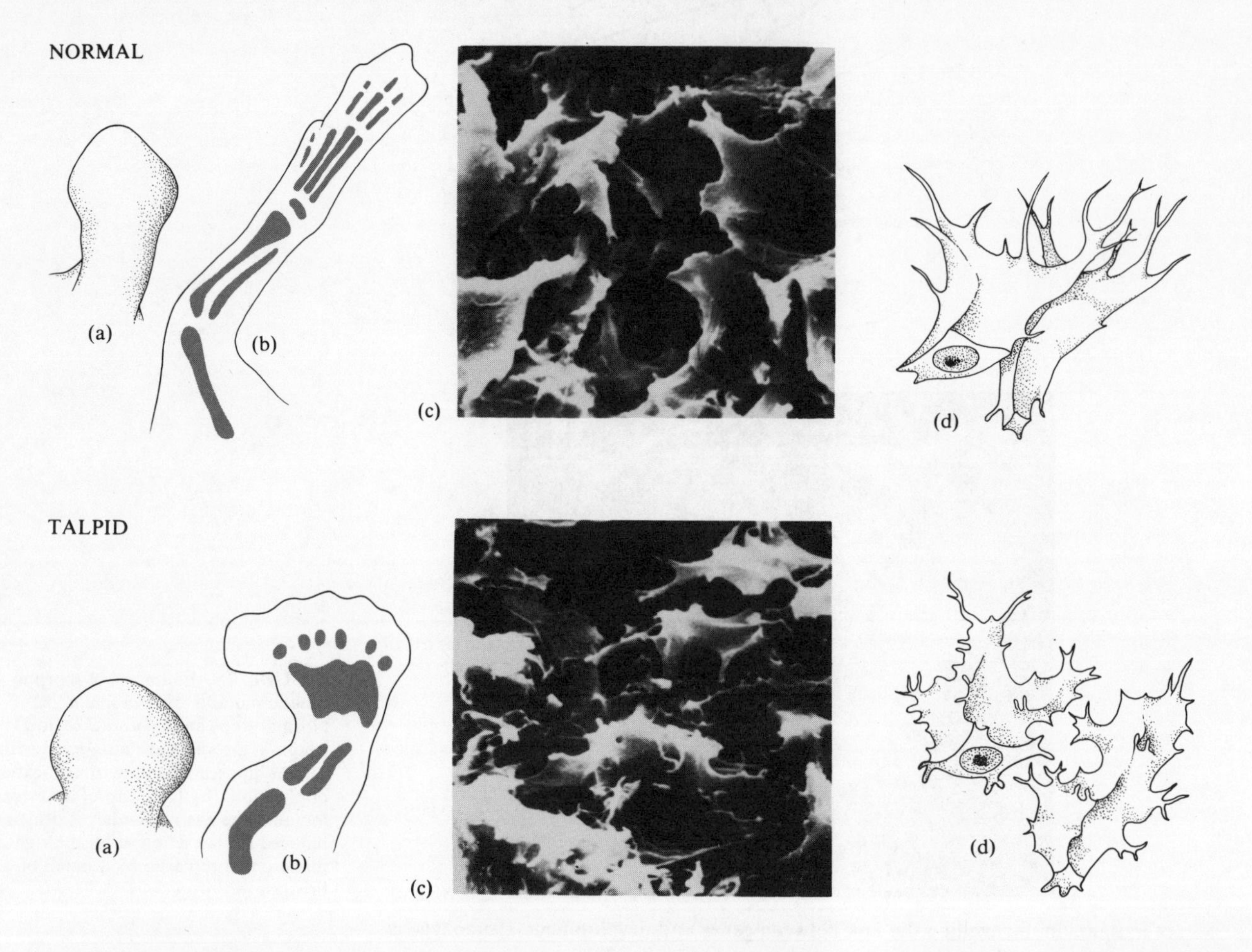

FIGURE 46 Normal (top) limbs and limb mesoderm cells and *talpid* (bottom) limbs and cells. (a) Embryonic limbs. (b) Limbs with skeletal elements shown. (c) Scanning electron microscope pictures of normal and *talpid* limb mesoderm cells. Note the cytoplasmic processes of *talpid* cells. (d) An impression of the shape of individual cells in transverse section.

embryonic limb cells are separated and cultured on a plastic dish, they move around before settling down to form a fused network. The cells are elongated, with a ruffled membrane at the leading edge (see Section 4). *Talpid* cells on the other hand are flatter, with cytoplasmic extensions all around the cell that attach to the plastic dish in culture and prevent the cells moving around. It was later shown in the developing limb itself that normal and *talpid* cells show the same differences in morphology as are seen when the cells are dissociated.

□ Can you suggest from this information why *talpid* limb rudiments are an abnormal shape?

■ Possibly the impairment of cell movement in *talpid* cells prevents normal morphogenesis of the limb: the normal limb is long and thin, and cell movement may be necessary to produce this elongation.

Talpid shows both how simple cell properties can be affected by mutation and how these properties *might* affect morphogenesis. We emphasize the word 'might', for the complex interactions cells go through to build new shapes are not well understood. It is possible to use genetics directly to look at cell behaviour *in vivo* however, and this will be discussed next.

9.3 Cells and their lineages in development

You read earlier how simple cell properties may be important in producing changes in shapes in tissues, for example in sea-urchin gastrulation, and how mutants that affect particular cell properties such as cell mobility may also affect morphogenesis (Section 9.2). So far, we have not really considered an important feature of development: cell division. Cells in development are, by way of growth and division, constantly increasing in number, and they all bear, to a lesser or greater extent, a kinship with one another. All are derived from the fertilized egg, but in a structure such as an imaginal disc, it may be 10 or 11 cell divisions since cells on opposite perimeters of the disc had a common parent cell.

We have considered genes as acting on all cells of an organism. A mutation that prevented glycolysis, for example, would be lethal, because no cell could exist without this essential process. A mutation affecting the production of eye pigment would only be *expressed* in the eye: although the eye pigment genes are present in every cell, they are only active in the eye pigment cells.

This raises an important question. Are cells differentiated because of *where they are* or because of their *ancestry* (lineage)?

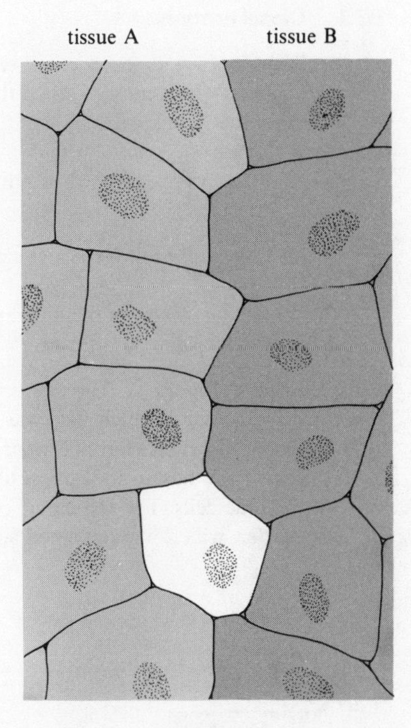

FIGURE 47

□ Look at Figure 47, which shows cells at the boundary of two embryonic tissues A (*pink*) and B (*grey*). The white cell has recently been formed from a type A parent cell. Will it become an A cell or a B cell if (a) it is differentiated according to its position, or if (b) it is differentiated according to its ancestry.

■ To be differentiated by position, the cell must be capable of receiving some kind of signal from its neighbouring cells and acting on this information. Depending on the signal, the cell would then differentiate. (b) Differentiation by ancestry suggests that the type of cell is determined by its parentage. The cell would therefore become an A cell.

The total family of cells derived from a single ancestor cell by mitosis is called the *clone* produced by this ancestor. Thus an organism is, strictly speaking, the clone produced from a fertilized egg. In order to decide between the possibilities of differentiation by position or lineage, it is first necessary to mark the cells in some way so that those with common lineages can be observed. In the early days of experimental embryology, clones were observed during development by labelling cells with fine carbon particles or dyes. After cell division, the ingested particles were seen in the two daughter cells, and the growth of the clone during subsequent cell divisions could be observed. In this way, it was possible to construct fate maps (see Figure 48) showing which parts of embryos form which adult structures.

clone*

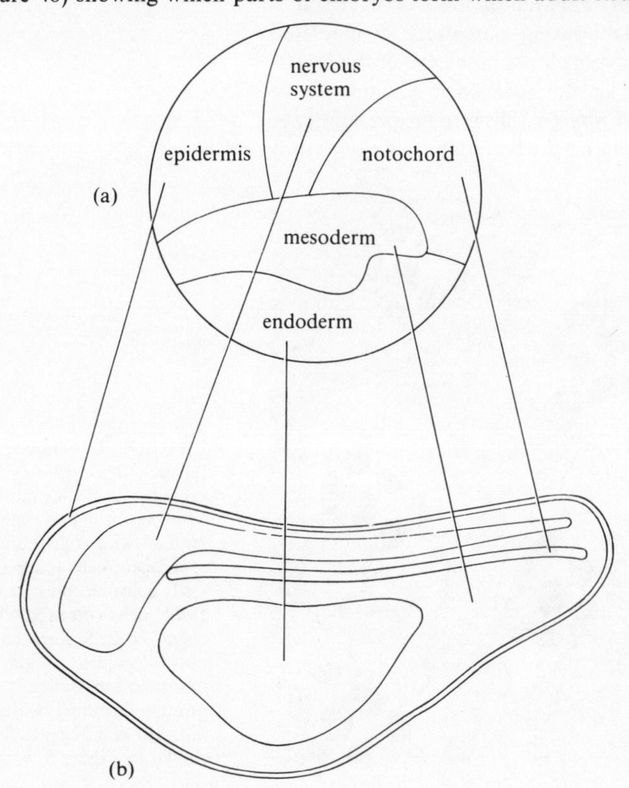

FIGURE 48 The construction of a fate map in an amphibian. Regions of the egg (a) are labelled with carbon particles and they become localized to particular embryo regions (b)—highly simplified. Each part of the egg, therefore, is destined to form a particular part of the whole organism.

9.3.1 Clonal analysis

What does a marked clone tell us about the differentiated state of the cells it contains? Imagine a 'French flag' (see Section 2.3) made up of cells of red, white and blue types.

□ What happens if these differentiated cells go through a number of cell divisions?

■ They will each form a clone.

Do the clones carry the differentiated state of the original red, white or blue cells from which they are formed?

Look at Figure 49. Three cells were marked (by some hypothetical mechanism) at the time the 'French flag' was formed (a). After several cell divisions, the flag is larger and clones have been formed by the marked cells (b). Now consider (c). The blue clone has straddled the blue/white border, as have (presumably) the rest of the blue cells. The red clone, on the other hand, has stayed within the white/red border, which has remained straight.

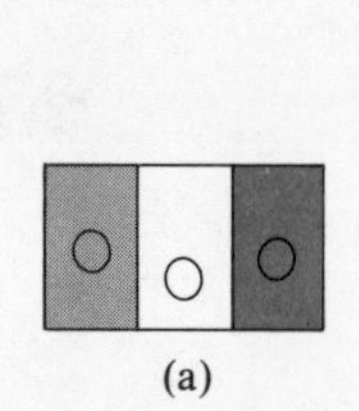
(a)

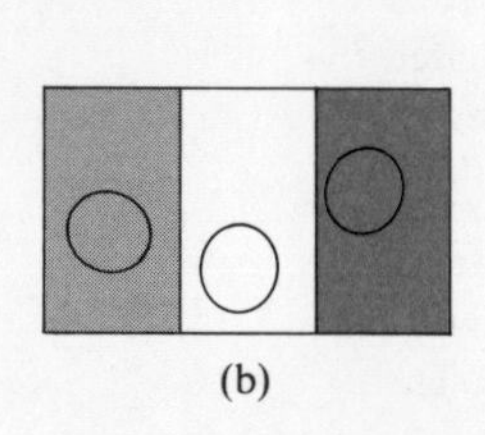
(b)

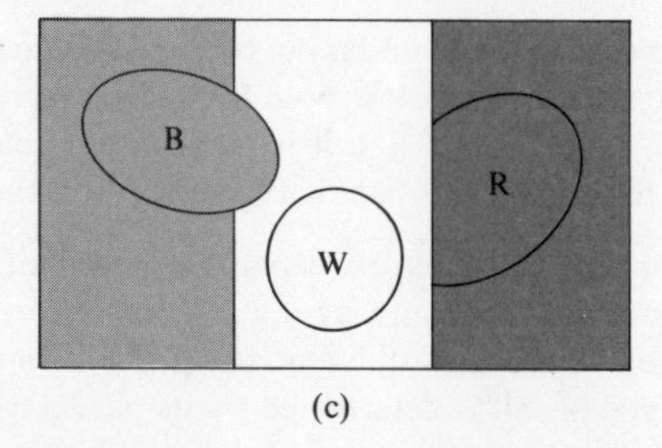

(c)

FIGURE 49 Clone growth in a cellular 'French flag'. The blue (grey) clone crosses the blue/white border, while the red clone respects the border.

If the clones maintain their initial differentiated state, all the blue clone cells will be blue: even the part that overlaps the white region. The red clone, because it stays *within* the red region, remains red. If the cells differentiate by *position*, all blue cells in the white region will become re-differentiated as white cells.

clonal analysis*

So, *clonal analysis* can help us distinguish between lineage-dependent and position-dependent differentiation. Consider the following example. The individual elements of the insect compound eye are called *ommatidia*—each element is made up of a group of pigment cells, light receptor cells and corneal cells. When the surface of the compound eye is observed, the part of the ommatidium seen is the overlying corneal facet (see Figure 1). Is an ommatidium made up of cells that share the same immediate ancestry, or do neighbouring ommatidia 'steal' related cells? This question may be answered by studying clones in the eye of the insect: this can be done in *Drosophila*, *Oncopeltus* or the cockroach. Clones can be marked by X-rays, or by introducing a graft of 'foreign' cells into the pre-adult eye to be studied. Figure 50 shows that ommatidia on the boundaries of clones may

ommatiduum (*pl.* ommatidia)

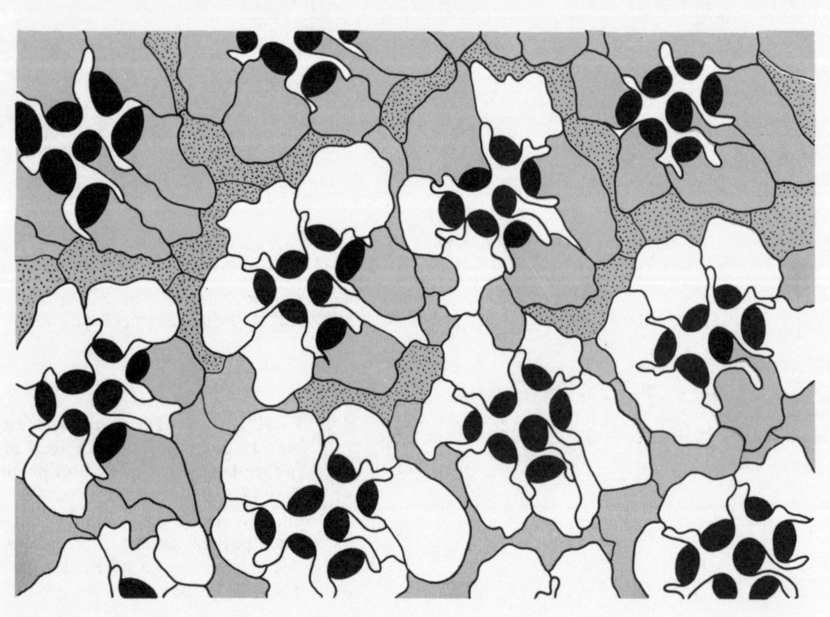

FIGURE 50 Transverse section through *Drosophila* eye showing several ommatidia. Each ommatidium is made up of seven light-sensitive cells (plain pink or white) surrounded by a number of pigment cells (stippled pink or grey). Cells coloured pink are part of a marked clone; cells coloured white are normal cells. It can be seen that the clone does not follow the boundary of each ommatidium, but cells within each ommatidium can be derived from different cell lineages (in the manner shown in Figure 51).

be constructed from cells not derived from the same immediate parent. Clearly, all cells are related to some extent: they all come from the single-celled egg! Figure 51 demonstrates how the lineages of lineage-dependent and position-dependent ommatidial formation would differ.

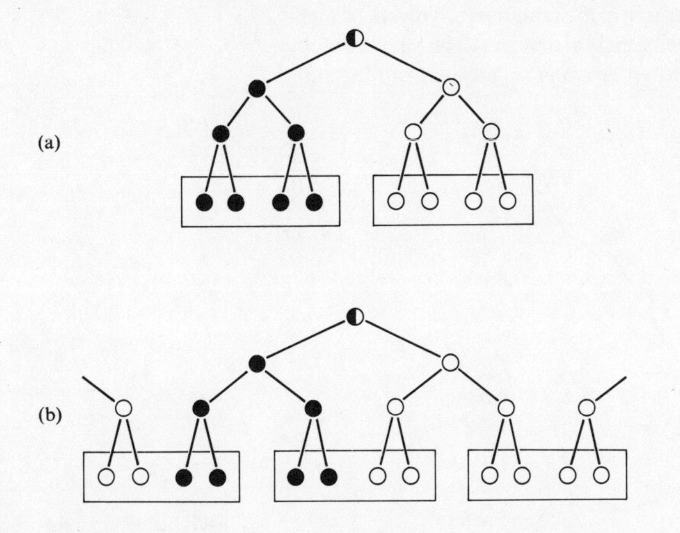

FIGURE 51 Cell lineage and ommatidial formation. In (a), the lineages of the two ommatidia are distinct, whereas in (b), there is overlap. (a) is lineage-dependent while (b) is position-dependent ommatidial formation.

9.3.2 The uses of clonal analysis

Although for technical reasons, clonal analysis cannot be used on all organisms, we have already seen that position alone is often not sufficient to guarantee the acquisition of a particular differentiated state by a cell. The grafting experiments performed on embryos (Section 8) demonstrated that cells must be competent to respond to signals from the surrounding tissue. There are few instances where lineage-dependent mechanisms have been shown to act. We have already considered one example in Section 2—the formation of two types of daughter cell by *Anabaena* cells, only one of which may form a heterocyst. Another example is seen in the development of the nematode worm species *Caenorhabditis* where the whole adult (consisting of only a few hundred somatic cells) develops as a result of a rigidly defined lineage.

We now examine a type of lineage-dependent mechanism recently demonstrated in *Drosophila* by the use of clonal analysis.

9.3.3 Compartmentation in development

Is there some kind of genetic switching mechanism that restricts clones of cells to certain disc regions (for example, the red clone in Figure 49)? If such a mechanism was acting, at some time in development, cells would become restricted in such a way that they would not cross genetically defined boundary lines in the disc. Such lines are called *compartment boundaries*, and cells either side are in *compartments*.

compartment boundaries
compartments*

This would mean that clones induced after a particular time in development would respect borders and would not cross them. Although data obtained by making clones in *Drosophila* organs suggested that this was so, the small sizes of clones that could be produced in structures such as the wing or leg were not sufficient to resolve the problem, because they have no long enough boundary to mark such a line.

The answer to whether or not compartments existed came from a group of mutants called *Minutes*. These are dominant mutations, lethal in homozygous form, that increase the time of normal larval development to about one and a half times the normal length. Several different *Minute* mutations exist, and they may be on different *Drosophila* chromosomes. By using appropriate genetic crosses and X-rays, it is possible to produce normal clones not carrying the *Minute* gene in a fly that *is Minute*. In effect, producing such a clone means unleashing fast-growing cells in a slow-growing background. Garcia-Bellido's group at the University of Madrid did this work in 1973, analysing patterns of clones formed on the *Drosophila* wing. They found that clones produced at any time in development would not cross a line separating anterior and posterior parts of the wing. At later stages

***Minute* mutant**

of clone initiation, other demarcation lines were seen (Figure 52). First, cells would not cross between the part of the adult exoskeleton to which the wing is attached (which is also formed by the wing disc) and the wing. Clones induced even later in development would not cross between the dorsal and ventral surfaces of the wing. Figure 52 shows this progressive compartmentation. The results were especially surprising because of the physical area of the boundary involved. It might have been expected that given the extra developmental time available, the normally growing clone cells would overgrow their slower growing neighbours, but this only seems to happen *within* each compartment.

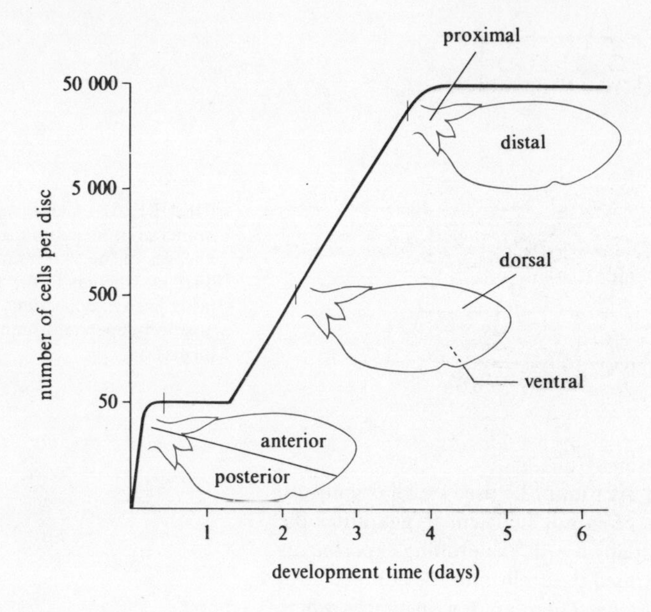

FIGURE 52 Progressive compartmentation in the *Drosophila* wing. Initial subdivision into anterior and posterior compartments is followed by segregation of dorsal and ventral tissue; later growth gives proximal and distal compartments. Cells from the various compartments do not cross compartment boundaries after they are formed.

As the cells in each clone are related, there must be a *genetic* switch that programmes the cells to keep within each compartment. What is the genetic basis of the boundary maintenance mechanism? At present we have little idea. A possibility is that cells in each compartment have some kind of label on the cell surface that prevents cells from different compartments from mixing. One theory is that similar mechanisms are in operation to those present in the cellular reaggregation system that we have already discussed in Section 5.1.

Extra evidence for compartmentation comes from studies on the segment borders of the milkweed bug *Oncopeltus* by Lawrence. He found that if clones of cells were marked by X-rays in a similar manner to the *Drosophila* experiments described earlier, then lines of clonal restriction can be seen here also. Each *Oncopeltus* abdominal segment was arbitrarily split into four quadrants. It was found that clones produced very early in development occupied all the quadrants, but those produced later occupied progressively fewer quadrants. In addition, clones never crossed between segments.

Several examples of clones in *Oncopeltus* are shown in Figure 53. There is a marked difference between the behaviour of clones at the segment border and those within each segment: the border cells are flattened, and seem to divide preferentially in an anterio–posterior direction, while internal cells divide randomly.

FIGURE 53 (a) A clone at an *Oncopeltus* segment border or margin (sm). Note the flattened cells and white band (wb) at the posterior of the leading segment. (b) An internal clone of cells showing the coherent shape of the clone.

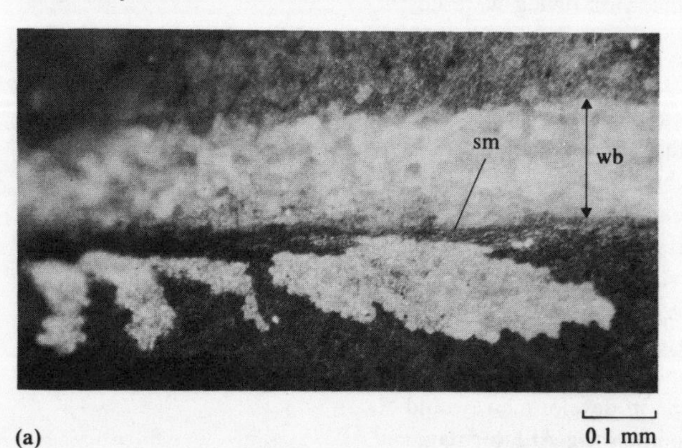

(a)

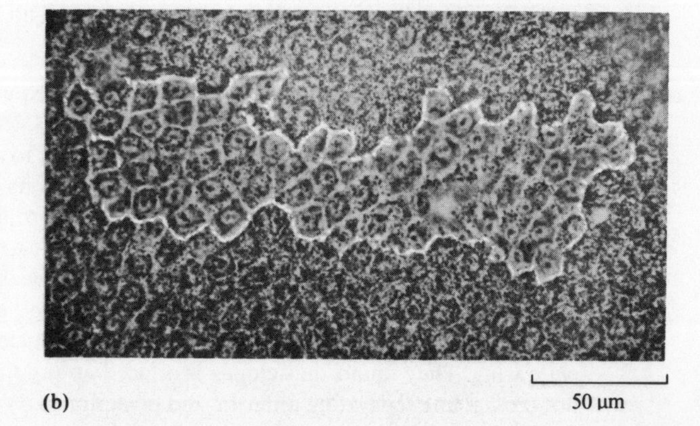

(b)

Whether or not this is true genetic compartmentation as demonstrated in *Drosophila* is open to debate, but the *Oncopeltus* findings certainly show that cells can be held in 'strictly disciplined' regional groups. An extra piece of evidence is that cells become flattened along the segment borders due to some cell recognition effect preventing the mixing of cells from different segments. Direct observations on *Oncopeltus* eggs has shown that the acquisition of independent lineages of segment cells (determination) in the presumptive epidermis certainly precedes any outward signs of segmentation (differentiation).

9.3.4 Chimaeric mice

There is a second way in which organisms of mixed genotype can be produced, and that is to combine cells from two embryos of different parentage during blastula formation (when each consists of only 4 or 8 cells). This is usually done with mouse embryos, although the technique has been used with other small mammals. Such a composite animal is called a *chimaera* and can develop into an adult.

chimaera*

This procedure is not as versatile as clonal analysis in insects. Instead of one cell being marked and consequently forming a clone, a half of each adult will be clonally derived from the half of the embryo that gave rise to it.

□ Why might this prevent analysis of the sort possible with *Drosophila* clones?

■ Clonal analysis in *Drosophila* involves marking single cells that will occupy a part of an adult organ: in this way it can be more 'sensitive' than the gross marking of a half of the embryo. In addition, chimaeras can only be formed at a particular time in development, so progressive clonal restrictions could not be observed. Finally, the *Minute* technique used so successfully in *Drosophila* cannot be applied to mammals.

Mammalian clones do not form integral single patches as do insects. By using appropriate genetic crosses, it is possible to produce *Drosophila* that are clonally derived from two differently marked cells. In insects of this type, there is always a sharp dividing line between the two groups of cells. In mammals, this does not occur: the two halves of the chimaera tend to intermingle, so that a hotchpotch of mixed tissue is seen in the adult. This basic difference is shown in Figure 54. It seems that mammal cells are capable of migration or perhaps selective 'aggregation' of particular types of cell, so that a clone of cells does not remain as a contiguous cell mass.

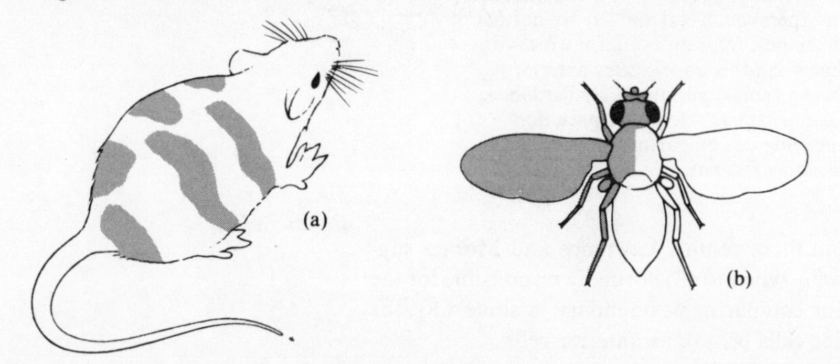

FIGURE 54 (a) A mouse produced by combining two embryos, one of which had lighter hair. (b) A *Drosophila* derived from a 'dark' and a 'light' cell. Note that the fly shows a smooth border between the two tissue types. In the mouse, there are patches.

Chimaeric mice have been used to reconstruct the origins and clonal ancestry of cells that give rise to the mouse coat. In a similar way, the fusion of two blastulae, one of which has a mutant enzyme activity that shows up with staining, has allowed clonal relationships in internal organs to be studied.

9.4 The genetic control of pattern formation

We have looked at the concept of compartmentation and the methods of clonal analysis that led to its discovery. The next step is to ask if any genes are known that might be implicated in the setting up and maintenance of compartment borders. The fact that compartments are populated by clonally related cells indicates that they may have a common differentiated state, and this poses additional questions. Why do compartments exist? Do they merely perform some kind of simple 'demarcation' role to keep development proceeding smoothly, rather like shop stewards in a factory who 'control' their workers and prevent them crossing the shop floor to do work outside their job description? Or is their role more important: do compartments represent *functional* subdivisions of tissues and organs?

These questions will now be at least partially answered, and we begin by looking at the one gene known to date that seems to affect the maintenance of the compartment borders.

9.4.1 The *engrailed* gene

Engrailed is a recessive gene in *Drosophila* whose mutant allele causes mirror-image duplication of anterior wing structures in place of the normal structures in the posterior of the wing. By making clones of *engrailed* cells in a non-mutant fly, and by marking the clones with a suitable bristle marker so that they would be visible on the adult wing, Lawrence and Morata were able to show that clones in the anterior of the wing behaved normally, while those in the posterior produced a distorted wing, the clones distinctly crossing the anterior–posterior compartment boundary (Figure 55). Non-mutant type clones induced in an engrailed

engrailed* gene

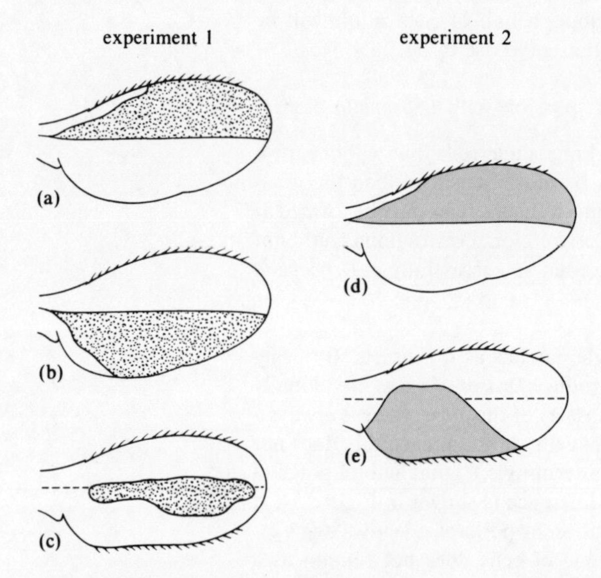

FIGURE 55 Examples of clones produced in two experiments with the gene *engrailed*. In experiment 1, (a) and (b) are normal wings with marked normal clones; (c) is an *engrailed* wing with a normal clone. In experiment 2, (d) is an *engrailed* anterior clone in a non-*engrailed* background, and (e) is a similar clone, this time in the posterior compartment. Note that posterior *engrailed* clones crossed the anterior compartment boundary, whereas anterior *engrailed* clones did not.

background behaved normally. From these results, Lawrence and Morata suggested that the *engrailed* gene in its wild-type form is normally responsible for the maintenance of the anterior–posterior compartment boundary in some way, the mutant effect being to make posterior cells behave as anterior cells.

The *engrailed* gene is of interest to developmental biologists for several reasons. First, it is a gene that in the mutant state, produces a deformity in wing shapes and thus shows itself to be important in morphogenesis. Second, this deformity is *cell autonomous*, that is it is only expressed by the cells with the mutant gene: this is shown up by clonal analysis, because only homozygous *engrailed* cells show an abnormally shaped wing area. Finally, as mentioned above, the wild-type *engrailed* gene seems to be important in the maintenance of a developmental restriction line over which cells cannot cross during development.

cell autonomous mutant*

9.4.2 Homeotic genes

The wild type allele of the *engrailed* gene thus seems to be involved in the maintenance of a compartment boundary. This boundary, however, separates two regions of the *Drosophila* wing with no obvious functional difference. *Engrailed* could therefore be used as evidence that compartments perform the mere 'housekeeping' function of maintaining proportion and regionalization in development.

Is there evidence that compartments can represent *functionally* different regions? In Section 9.1, we described a group of mutants called homeotics that substitute parts of *Drosophila* organs one with another. An example is the mutant *antennapedia*, which partially transforms the antenna into a leg. Our knowledge of the existence of compartments has led to speculation that perhaps these homeotic changes are compartment specific. There is a group of homeotic mutants that lend support to this idea: the *bithorax* series, described and studied by E. B. Lewis.

The fly embryo is divided into a series of segments from which appendages develop. These segments broadly fall into three groups, head, thorax, and abdomen. The wings and associated body parts form from the *meso*thoracic segment, while the halteres form from the *meta*thoracic segment. Genes of the *bithorax* series transform parts of these segments to those formed normally by other segments. Now study Table 2 and Figure 56.

***bithorax* genes**

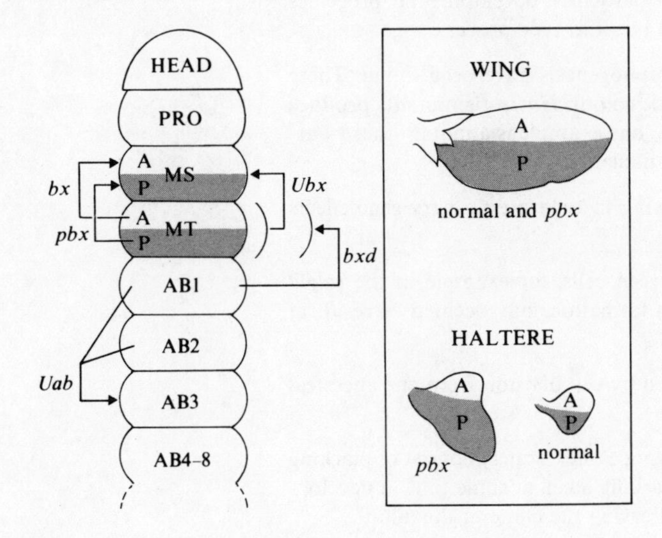

FIGURE 56 The action of some of the *bithorax* series of genes in *Drosophila*. For simplicity, the segments have been shown as they would appear in the larva. Inset is the appearance of the haltere and wing in *pbx* flies where the posterior haltere compartment becomes posterior wing tissue. Arrows show the direction of transformations; for example, *pbx* results in the change posterior MT to posterior MS.

TABLE 2 Effects shown in Figure 56

Gene	Gene symbol	Changes ···	To ···
bithorax	(*bx*)	anterior metathorax (AMT) (anterior part of haltere)	anterior mesothorax (AMS) (anterior part of wing)
postbithorax	(*pbx*)	posterior metathorax (PMT) (posterior part of haltere)	posterior mesothorax (PMS) (posterior part of wing)
Ultrabithorax	(*Ubx*)	metathorax (MT) (haltere)	mesothorax (MS) (wing)
bithoraxoid	(*bxd*)	1st abdominal segment (AB1)	metathorax (haltere)
Ultra-abdominal	(*Uab*)	1st, 2nd abdominal segment (AB1, AB2)	3rd abdominal segment (AB3)

The inset to Figure 56 shows a central finding that has emerged from studies of these mutants: transformations do not occur only between whole segments, for example AB1 to MT, but also between parts of segments. Not only this, but the transformations may be compartment specific. In *postbithorax*, posterior haltere becomes posterior wing. In *bithorax*, anterior haltere becomes anterior wing.

□ A fly with the double genotype *pbx bx* is produced. What abnormality might you observe?

■ The fly will have two sets of wings, because both anterior and posterior haltere compartments will be transformed into the corresponding wing compartments.

These studies show that functional differences may be related to compartments. As we have seen, presumptive haltere cells may become wing cells in a compartment-specific manner. It seems that in *Drosophila* at least, development proceeds by progressive differentiation, which switches on whole groups of cells that need not be directly clonally related. As you will see in the next Unit, there is some evidence that gradients of positional information are present in eggs, and it has been suggested that the initial segmentation patterns in *Drosophila* are set up in this way. The *bithorax* system might therefore represent gene defects that change the interpretation of the positional information.

Summary of Section 9

1 The analysis of mutant genes affecting particular developmental processes may give insight into the genetic control of the wild-type process.

2 Many *Drosophila* mutants affecting morphogenesis have been found. These include genes affecting wing shape and bristle colour. *Homeotic* mutants produce adult organs from the 'wrong' imaginal disc: one example is a mutant that transforms part of the compound eye into wing tissue.

3 Analysis of gene action may involve trying to separate primary gene effects from secondary 'pleiotropic' effects.

4 Genes may affect the interactions between cells, for example in the *talpid* mutant of the chick where abnormal limb formation may occur as a result of reduced cell mobility.

5 A clone is the total group of cells formed by cell division from one ancestral cell.

6 Clonal analysis is a technique for observing clones. It may consist of marking cells with carbon particles or genetically marking a cell at some time in development (for example, by X-rays) and then observing the clone in the adult.

7 Clonal analysis can be used to find out if organs are formed as a result of the cells in a particular *position* being differentiated or if cells of a particular *ancestry* become differentiated.

8 A genetic compartment is a region that limits a clone during development so that cells from one clone do not cross over into other delimited regions. Such compartments have been found in *Drosophila* and possibly also in the bug *Oncopeltus*.

9 Chimaeras made by mixing cells from two genetically distinct early mouse blastulae are used to study cell lineage in mammals.

10 A basic difference between insect and mammalian clones is that related cells in mammals are more likely to separate during development.

11 Several *Drosophila* genes are known that seem to be compartment specific. These include *engrailed* and some genes of the *bithorax* system.

12 The wild type allele of the *engrailed* gene is involved in the maintenance of the antero-posterior border in the wing.

13 *Bithorax* genes control the fate of some segments in *Drosophila.*

14 The mechanism of compartmentation has been suggested to be a gradient-induced pattern specified early in development.

Objectives and SAQs for Section 9

Now that you have completed this Section, you should be able to:

★ explain how the study of mutants may contribute to the understanding of normal development.

- ★ outline how clonal analysis can help decide whether organs form by position-dependent or ancestor-dependent mechanisms.
- ★ outline the experimental evidence for the phenomenon of compartmentation.
- ★ describe the evidence that shows how the wild type allele of the *engrailed* gene may be involved in the maintenance of compartment borders.
- ★ outline how the *bithorax* system shows that compartments may be segment specific, and how genes may control segmentation in *Drosophila.*

To test your understanding of this Section, try the following SAQs.

SAQ 20 (*Objective 16*) A developmental biologist finds five mutants (i)–(v) of *Drosophila.* Which mutant would he study if he wanted to investigate the following problems (a)–(d)? (Some may fit into more than one category.)

(a) The development of the *Drosophila* eye.

(b) Pattern formation.

(c) Homeotic mutation.

(d) The differentiation of pigment cells.

The mutants are as follows:

(i) A mutant gene that produces specific eye colour in the wild-type but none in the mutant.

(ii) A mutant gene that produces a lighter body colour than the normal allele of the same gene.

(iii) A mutant gene that transforms antennal tissue into leg tissue.

(iv) A mutant gene that produces eyeless flies.

(v) A mutant gene that changes the position of veins on the wing.

SAQ 21 (*Objectives 17 and 18*) Are the following statements true or false?

(a) *Minute* mutants are mutants that develop more quickly than wild-type flies.

(b) The first compartment boundary seen in the *Drosophila* wing is the antero-posterior one.

(c) The evidence from *Oncopeltus* suggests that compartments exist.

(d) Chimaeric mice can be produced at any stage during development.

(e) Cells in *Drosophila* clones tend to separate during development.

(f) *Oncopeltus* clones at segment borders appear different from clones within each segment.

SAQ 22 (*Objective 19*) A biologist wants to test the hypothesis that the *engrailed* gene can affect compartment boundaries in other *Drosophila* organs besides the wing. He finds that a compartment border is present in the antenna, and does experiments to test this hypothesis in a similar manner to those described in Section 9.4.1. Which of the following observations would be in agreement with his hypothesis?

(a) Normal clones in a normal wing cross the compartment boundary.

(b) *Engrailed* clones in the posterior of the normal antenna cross the compartment boundary.

(c) *Engrailed* clones in the anterior of the normal antenna cross the compartment boundary.

(d) Normal clones in an *engrailed* antenna cross the compartment border.

SAQ 23 (*Objective 20*) It has been suggested that *Drosophila* evolved from ancestors with two pairs of wings. How does the presence of the *bithorax* series of genes support this view?

10 Summary and conclusions to Unit 14

One of the problems of genetic analysis is that it is largely limited to organisms with fast generation times such as *Drosophila*, and it is a big leap from a fly to man. Many developmentalists mistrust generalizations based on work with insects: do compartments also exist in man?

In this Unit, we have spanned the whole area of pattern specification, morphogenesis, and the genetic control of these events. We started by considering the underlying 'signals' or instructions that define the spatial organization of patterns and shapes in development. Cells may acquire positional information by 'reading' the local concentration of a gradient substance and interpret this value to specify their molecular differentiation (TV programme: *Patterns in Development: Gradients*).

There is little convincing experimental data to support this, although there is 'circumstantial' evidence from several systems: regeneration and wound healing experiments especially. The model really poses many more questions than it answers—how are gradients established and maintained? How are the characteristic boundary values of the gradient substance established? How can cells measure the local concentration of the gradient substance? What mechanism allows the appropriate selection of genes to turn on according to the position of the cell in the gradient?

Next, we looked at the cellular properties involved in the origin of embryonic shape and form. We started by considering the kinds of properties of cells *in vitro* that might give clues as to how more complex *in vivo* cell properties might affect development. The phenomenon of cellular reaggregation, shown in the TV programme *Cell Movements*, suggests that cells have the ability to move actively toward cells of their own type. Cells therefore seem to take an active part in development, and do not only act as 'sheep' 'shepherded' by positional information rules.

The story of shape generation was continued by considering the process of gastrulation in the sea-urchin and the amphibian. Here we saw how cell properties such as adhesiveness and movement are important in morphogenesis. Cells may also respond to instructions from their neighbours, and Section 8 dealt with this phenomenon of embryonic induction.

We concluded the Unit by looking at how genes control development. Our understanding of genetic control processes is still rather limited, but the use of techniques such as clonal analysis, which allows the behaviour of a genetically distinct 'island' of cells to be observed, is being used increasingly in research.

Objectives for Unit 14

Now that you have completed this Unit, you should be able to:

1 Define and use, or recognize definitions and applications of, the terms marked by an asterisk in Table A.

2 Describe how a gradient might specify positional information and so help to form and maintain a developmental pattern. (*SAQs 1, 2 and 4*)

3 Define positional information using the French flag model; discuss how the model applies to two types of regeneration. (*SAQs 1–3*)

4 Distinguish between mosaic and regulative development. (*SAQ 2*)

5 Describe the evidence for and the possible mechanism of an insect segmental gradient. (*SAQ 3 and 4*)

6 Summarize the important features of pattern formation in *Anabaena*.

7 Briefly describe the life cycle of *Drosophila*, and give reasons for the usefulness of *Drosophila* as an organism for developmental studies. Discuss why the structure and biology of imaginal discs make them useful models for pattern formation. (*SAQs 5–7*)

8 Describe how regeneration and duplication experiments in *Drosophila* imaginal discs give evidence for a gradient of positional information. (*SAQs 5–7*)

9 Describe how fibroblasts and other specified cells move and are oriented *in vitro*. (*SAQs 8 and 9*)

10 List the properties of single cells that may be important in morphogenesis. (*SAQ 9*)

11 Describe critically the various hypotheses proposed to account for sorting out. (*SAQs 10 and 11*)

12 Briefly summarize gastrulation in the amphibian and sea-urchin embryos and explain how cell properties such as adhesiveness and movement may be important in these processes. (*SAQs 12–14*)

13 List the differences between gastrulation in simple and complex animals by comparing the sea-urchin and amphibian. (*SAQ 15*)

14 Summarize experimental approaches used in attempts to characterize the inductor and discuss the regional specificity of inductive effects. (*SAQs 16–18*)

15 Summarize how ectoderm–mesoderm interactions provide a model for studying cell induction. (*SAQ 19*)

16 Explain how the study of mutants may contribute to the understanding of normal development. (*SAQ 20*)

17 Outline how clonal analysis can help decide whether organs form by position-dependent or ancestor-dependent mechanisms. (*SAQ 21*)

18 Outline the experimental evidence for the phenomenon of compartmentation. (*SAQ 21*)

19 Describe the evidence that shows how the wild type allele of the *engrailed* gene may be involved in the maintenance of compartment borders. (*SAQ 22*)

20 Outline how the *bithorax* system shows that compartments may be segment specific, and how genes may control segmentation in *Drosophila*. (*SAQ 23*)

ITQ answers and comments

ITQ 1 Migrations of cells into the blastocoel, folding inwards of cell sheets, followed later by extension of pseudopodia and the pulling of archenteron towards the animal pole.

ITQ 2 Primary mesenchyme; cells with pseudopodia.

ITQ 3 Formation of a blastopore; invagination of cells to give an archenteron; changes in cell shape around the blastopore.

ITQ 4 (a) The surface layer and the bottle cells appear to be responsible for blastopore formation. The bottle cells move inwards and yet retain a connection to the surface layer by a long tapering 'neck' (see Figure 38). The tightly coherent surface layer is dragged inwards to form a blastopore groove. (b) The adhesive properties of the surface layer are obviously important, both in the formation of the blastopore and in the coordination of the various events at gastrulation. The spreading of the mesoderm beneath the ectoderm depends on the adhesive properties of the two cell layers. (c) (i) The inherent tendency of certain parts of the gastrula surface to stretch; (ii) the spreading of the mesodermal cells of the archenteron roof. (d) (i) The coordination of the various events at gastrulation; (ii) the production of a coherent cell layer that plays a part in blastopore formation. (e) (i) The darkly pigmented ectoderm spreads during gastrulation; (ii) the mesoderm of the archenteron roof spreads beneath the ectoderm.

SAQ answers and comments

SAQ 1 (a) True. (b) False: if this were true, all cells would have the same positional information. (c) False: it can account for both regulation by growth and for regeneration by remodelling. (d) True. (e) False: existing cells have their positional values reassigned. (f) True: you would need a two-dimensional gradient to generate a Union Jack.

SAQ 2 (a) This is done by new cells being produced at the cut edge. As low gradient values are assumed to be missing, diffusion of morphogen from high to low concentration will occur into the new cell region. (b) The gradient here becomes steeper, possibly by a new sink being re-established at the cut edge.

SAQ 3 A distortion of the ripple pattern occurs in (b), (c), (e) and (g). Recall that a disturbed ripple pattern is thought to arise when cells at different gradient levels are adjacent. So, if integument (cuticle) pieces are placed at an identical level in the same segment in the same orientation (a and d), the adult ripple pattern is normal. Similarly, transplantation to a site at the same level in an adjacent segment produces a normal ripple pattern (f). If the pieces are rotated or are transplanted to a different level in the *same* segment, the ripple pattern is abnormal (b and c). Pieces that are *rotated* and then transplanted to the same level in an adjacent segment cause an abnormal ripple pattern (e). Pieces transplanted into integument at a different level in an adjacent segment also produce a disturbed ripple pattern.

SAQ 4 Figure 16 clearly shows that some of the host cells (with their nuclei in black) have changed their developmental fate as a result of some influence from the graft. For instance, just behind the graft cells, type 1 cuticle has developed from host epidermal cells that would otherwise have produced type 2 cuticle. Similarly, in front of the graft, host cells have produced 6, 5, 4 and 3 cuticle from presumptive type 2 epidermal cells. What is the cause of these changes? If you are unable to answer, read Section 2.5 again, study Figure 9c and Figure 14 carefully, and then read (a) below, which hints at the full answer given in (b).

(a) Consider gradient levels either side of any of the thick vertical lines in Figure 9c. To the left, which we have called the posterior part of the segment margin, there is the lowest gradient level. To the right, which we have called the anterior part of the segment margin, there is the highest gradient level. A complete segment margin therefore displays two different gradient levels. So, region 0 of the graft, the posterior margin, has a low gradient position while region 6 of the graft, the anterior margin, has a high gradient position.

(b) If region 0 maintains a low gradient position, a local dip is produced in the gradient landscape following its implantation into presumptive region 2, which has a higher gradient value. This is an unstable situation, which causes flow of gradient material. This produces a change in the gradient level of the host presumptive type 2 cells behind the graft, which now adopt a level corresponding to presumptive type 1 cells. Similarly region 6 maintains a high gradient position, producing an unstable hill in the gradient landscape. So, a flow of gradient material occurs, and gradient levels corresponding to types 6, 5, 4 and 3 are established in the host cells in front of the graft.

So, the cuticular structures that develop can be explained if epidermal cells produce cuticle in accordance with their position in a gradient.

SAQ 5 The result would be basically the same as the 4, 5, 6 fragment in Figure 25. Disc parts from higher up the gradient may form those lower down, so 4 and 5 give 6, as well as duplicating themselves.

SAQ 6 W—Regeneration. This section is the upper medial quarter, which presents the high point of the gradient, as you can see in Figure 23b. This part of the disc has the property of regenerating all other parts. X—Duplication. This part is complementary to W, and as such does not contain the highest parts of the gradient. Therefore it is only capable of duplicating the structures present. Y—Regeneration. This part of the disc contains the upper medial quarter, and so is capable to regenerating all structures. Z—Duplication. This segment is complementary to Y, and so is only capable of duplication.

SAQ 7 (a) ii; (b) ii; (c) iii.

SAQ 8 (a) False. (b) True. (c) True: fibroblastic movement occurs only over a solid or semi-solid substratum. (d) True. (e) False: the inherently precise machine is used to explain fibroblast orientation.

SAQ 9 Cell movements may be oriented by the substratum—this is called *contact guidance*, and the substratum may aid differentiation, as in the case of muscle cell fusion to give muscle tubes. *Contact inhibition* results when the leading lamellae of fibroblasts in culture contact one another. The cells may also align, as in the case of the muscle cells and the fibroblasts studied by Elsdale.

SAQ 10 (a) Mesoderm cells. It is interesting to note that the 'external' cells of organisms, the ectoderm, tend to be on the periphery of mixed tissue reaggregates. It is perhaps to be expected that such cells have a special mechanism for maintaining such a position. (b) By suggesting that cells reacquire their abilities to adhere only after a time lag differing for each cell type. Cell type 1 would first reaggregate, cell type 2 would then form around cell type 1, and so on.

SAQ 11 B cells will form a ball inside A cells, because the stickiest cells take the central position. Because B is stickier than C, and C is stickier than A, B cells must be stickier than A cells.

SAQ 12 (ii), (iii) and (iv). The main features of primary invagination are the migration of vegetal cells into the blastocoel, and Gustafson and Wolpert suggest that changes in adhesiveness between vegetal cells cause archenteron formation by making the tissue sheet bend inwards.

SAQ 13 (a) The inward movement of surface cells (bottle cells). Presumably some kind of positional information inherent in the early embryo sets which region will form the blastopore. (b) It contracts as the archenteron increases in size and so 'squeezes' it. (See Figure 37.) (c) The presumptive ectoderm. (d) The archenteron. (e) Explanting endoderm tissue and putting a graft of presumptive bottle cells onto it. The blastopore cells sink into the endoderm forming a groove, and because they are darkly pigmented ectodermal cells, their progress can readily be followed.

SAQ 14 (a) iii; (b) i; (c) vi; (d) iv; (e) ii; (f) v. Moderate contact between the cells and between the cells and the supporting membrane (i) is shown in (b). The result of increased contact between the cells and reduced contact with the supporting membrane (iv) is shown in (d). The effect of the cells in (d) reducing their contact if the ends of the sheet are fixed is given in (a). (e) shows the result of a further increase in contact between the cells compared to that in (c) if the cells do not lose their contact with the supporting membrane (ii). The effect of an increased contact between the cells if the cells do not alter their contact with the supporting membrane (vi) is (c). (f) shows the effect of rounding-up the cells in (b) if the ends of the sheet are fixed (v).

SAQ 15 Similar processes occur in both cases, but the invagination step is slightly different. In vertebrates, specialized cells (bottle cells) are involved, while sea-urchin embryos rely on pseudopodia to 'pull' the archenteron into place.

SAQ 16 (a) False. (b) True. (c) False. (d) True. (e) True.

SAQ 17 (a) Normal. (b) Abnormal. (c) Normal. (d) Abnormal. (e) Non-inductor.

SAQ 18 The unheated bone-marrow extract produces only mesodermal inductions (e.g. muscles). So either mesodermalizing factor is present alone, or more likely, the effect of neuralizing factor is not evident. With increasing length of heat treatment, the induced structures tend more and more to show anterior and/or hind-brain characteristics. So, with increasing heat treatment, the effect of neuralizing factor becomes evident. Probably, this is due to the fact that mesodermalizing factor is destroyed by heat, while neuralizing factor is not (Section 8.2). The results in column B of Table 1 suggest that hind-brain characteristics are induced when the relative proportions of neuralizing and mesodermalizing factors reach a particular level. These results are in accordance with Saxen and Toivonen's hypothesis, which suggests that the type of structure induced depends on the relative proportions of two inductor substances, which differ in sensitivity to heat. Notice that prolonged heat treatment (150 seconds) appears to inactivate neuralizing factor, too.

SAQ 19 (a) True. (b) False. (c) True. (d) False. (e) True.

SAQ 20 (a) i, iv; (b) iii, iv, v; (c) iii; (d) i, ii.

SAQ 21 (a) False. (b) True. (c) True. (d) False. (e) True. (f) True.

SAQ 22 Observations (b) and (d).

SAQ 23 The presence of 'wing suppressing' genes would stop the metathoracic segment producing a set of wings. Halteres would be formed in their place. Thus the normal function of the *bithorax* genes might be 'wing suppression' and the mutant genes reduce or prevent the action of this suppression.

References and further reading

References to the Foundation Course

S101

1 Unit 18, *Natural Selection*, Section 6.

2 Unit 19, *Genetics and Variation*, Section 2.

S100

Unit 19, *Evolution by Natural Selection*, Section 1.

Unit 19, Section 1.

Further reading

If you have the time and interest to learn more about the topics discussed here, you may find the following relatively inexpensive books useful.

Ede, D. A. (1978) *An Introduction to Developmental Biology*, Blackie.

Garrod, D. (1973) *Cellular Development*, Chapman and Hall.

In addition, several more advanced textbooks have chapters that you may like to look at. Especially up-to-date examples are *The Developmental Biology of Plants and Animals* (1976), edited by C. F. Graham and P. F. Wareing, Blackwell, and *The Biology of Developing Systems*, by P. Grant, Holt Reinhart Winston (1977). Useful micrographs of developing embryos may be found in S. Mathews *Atlas of Descriptive Embryology*, Macmillan (1974).

Acknowledgements

Grateful acknowledgement is made to the following for permission to reproduce material in this Unit:

Figure 1d Bruce Coleman Ltd—photo by Grace Thompson; *Figure 5* Photo by Mike Wilcox, Cambridge; *Figures 6 and 7* L. Wolpert, *Current Topics in*

Developmental Biology, vol. 6, Academic Press Inc.; *Figure 8* Photo by Peter Lawrence, Cambridge; *Figures 9–11* P. A. Lawrence, Polarity and Patterns in the Post-Embryonic Development in Insects, in *Advances in Insect Physiology*, Academic Press Inc. (London) Ltd; *Figure 12* Lawrence, Crick and Munrow, Figure 3 in *Journal of Cell Science*, vol. 11, pp. 815–853, The Company of Biologists Limited, 1972; *Figure 16b* Davies & Balls (eds.), *Symposium* 25, The Society for Experimental Biology Symposia; *Figure 24* E. Hadorn, Transdetermination in Cells, Copyright © 1968 by Scientific American, Inc., all rights reserved; *Figure 28* D. R. Garrod, *Cellular Development*, Chapman and Hall Ltd; *Figure 29* P. A. Buckley and I. R. Konigsberg, Myogenic fusion and the duration of the post mitotic gap, in *Developmental Biology*, vol. 37, Academic Press, Inc., 1974; *Figure 30* T. Elsdale, Pattern formation and homeostasis, in G. E. W. Wolstenholme and J. Knight, *Ciba Foundation Symposium on Homeostatic Regulators*, J. and A. Churchill Ltd., 1969; *Figure 32* Steinberg, *Cell Membranes in Development*, Academic Press Inc.; *Figure 35* Gustafson and Wolpert, Cellular Movement and Contact in Sea-Urchin Morphogenesis, in *Biological Review*, Vol. 42(3), Cambridge University Press, 1967; *Figure 36* Gustafson and Wolpert, The Cellular Bases of Morphogenetics, in *International Review of Cytology*, vol. 15, Academic Press Inc., 1963; *Figure 38* J. Holtfreter, *Journal of Experimental Zoology*, vols 94 and 95, Wistar Institute of Anatomy and Biology Academic Press, Inc.; *Figures 40 and 41* L. Saxén and S. Toivonen, *Primary Embryonic Induction*, © Prentice-Hall Inc.; *Figure 42* L. Saxén and S. Toivonen, in *Annales Academiae Scientiarum Fennicae*, vol. IV (30), The Finnish Academy of Science and Letters, 1955; *Figure 44* L. Saxén and J. Wartiovaara, in *The Developmental Biology of Plants and Animals* (eds. C. F. Graham and P. F. Wareing), Blackwell, 1976; *Figure 46* D. A. Ede, *Introduction to Developmental Biology*, Blackie, 1978; *Figure 53* P. A. Lawrence, A clonal analysis of segment development, in *Journal of Embryol. Exp. Morph.* vol. 30, no. 3, The Company of Biologists Ltd., Cambridge University Press.

unit 15

Chicken or Egg?

Contents

Table A Scientific terms and principles used in Unit 15 2

Study guide for Unit 15 2

1 **Introduction** 2

2 **Early determination** 3
2.1 Mosaic development 3
2.2 Regulative development 5
2.3 Localization in regulative eggs 6
Objectives and SAQs for Section 2 7

3 **The plane of cleavage** 8
Summary of Section 3 9

4 **The nature of the localized factors** 9
Objective and SAQ for Sections 3 and 4 10

5 **Factors and cell differentiation** 10

6 **The origins of localization** 10
6.1 Polarity in the egg 11
SAQ for Sections 5 and 6 11

7 **Development: a summary** 12
7.1 From egg to adult 12
7.2 The questions that remain 12
7.3 Conclusions 12

Objectives for Unit 15 14

SAQ answers and comments 14

References to the Foundation Course 14

TABLE A Scientific terms and principles used in Unit 15

Assumed knowledge†	Introduced in an earlier Unit	Unit	Introduced or developed in this Unit	Page
cellular differentiation[2]	*Anabaena*	1, 14	epigenesis*	5
cytoplasm	animal/vegetal poles	11	gradients in egg*	9
monozygotic	animal–vegetal axis	11	localization*	4
mRNA (messenger RNA)[4]	blastomere	11	mosaic development*	4
transcription[3]	bilateral symmetry	1	polarity (of egg/embryo)*	11
translation[5]	cleavage	11	regulative development*	5
zygote[1]	competence	11	timing of determination*	9
	determination	11		
	differential gene transcription	12 & 13		
	diploid	11		
	embryonic induction	11		
	gradient	14		
	haploid	11		
	Jacob–Monod hypothesis	12 & 13		
	morphogenesis	11		
	pattern formation	14		
	polarity (of egg/embryo)	11		
	rhizoid	2		
	seaweed *Fucus*	2, 3		
	totipotent	12 & 13		

* These terms must be thoroughly understood—see Objective 1.

† Most of these terms are explained in the Science Foundation Course. Those that are of particular importance to the understanding of this text appear with a superscript number and have a full reference at the end of the Unit.

Study guide for Unit 15

The purpose of this Unit is two-fold. First, it is to introduce you to the idea that fertilized eggs or early embryos contain factors that influence the subsequent development of the organism. This topic forms the bulk of the Unit and the TV programme, *Chicken or Egg*? Second, the later part of the text should serve to summarize Units 11–15 as a whole.

The text is quite brief. However, we hope that this will give you an opportunity to revise some of the earlier Units in the *Development* Block; the summary and the questions that remain (Section 7) may help direct your attention to new ways of looking at earlier material.

1 Introduction

As you have seen from Units 11–14, development is about change. Cell numbers, the types of cell, the size and shape of the organism—all of these change during development. And all depend on an interaction between the genotype of the organism and its environment (Unit 11, Section 3.2). The genetic information in the initial zygote[1] is copied and partitioned among subsequent generations of cells. These cells interact with the surrounding environment and with each other to produce the subcellular changes characteristic of cellular differentiation[2] and the cell movements and divisions, that together lead to an adult multicellular organism. But how do these changes start?

It is relatively easy to understand that once an embryo contains cells of different types, then each cell is in a local environment that differs from the environment of any other cell. These different local environments can, in turn, produce further differential changes among certain groups of cells. Such local interactions are, of course, obvious in pattern formation as in *Anabaena* (Unit 14, and *The S202 Picture Book**), or the phenomenon of embryonic induction (Unit 14), for example.

* The Open University (1981) S202 PB *The S202 Picture Book*, The Open University Press. This contains colour and half-tone illustrations for the Course.

Once an embryo contains different local regions within it, these can, in turn, lead to further changes—the changes 'snowball' (Unit 11). But, to repeat the question, how do these changes start? At what stage in embryogenesis do the first local differences appear, and how? These are perhaps the key questions about development and it is fitting to end this treatment of the subject by attempting to answer them.

2 Early determination

As development in multicellular organisms is a cellular phenomenon (i.e. it depends on *cell* division, *cell* differentiation and *cell* movement), in attempting to answer the questions posed above we should begin by trying to define at what point in embryogenesis differences between cells can first be detected. One way is to ask: at what stage are individual cells first determined? That is, at what stage is the commitment of individual cells to their normal developmental fates irreversible?

This question occupied many of the great embryologists at the turn of the last century. The experimental strategy was simple and elegant. If any cell could be killed or removed from an early embryo without causing the later stages of the embryo to lack any tissues or organs, then the remaining cells in the embryo could not have been completely determined as they must have regulated their development to produce a normal later embryo. On the other hand, if the later embryo lacked any structures, then some determination must have already occurred by the time the cell was killed or removed.

2.1 Mosaic development

One such series of experiments was published by Conklin in 1906. The organism he studied was *Styela*, an ascidian (Urochordata; see *A Survey of Living Organisms**). He killed one of the blastomeres at the two-cell stage and examined the subsequent development of the other live blastomere. The sort of results he obtained are shown in Figure 1, in which the normal development of an intact two-cell embryo is compared with that of an embryo in which a blastomere was killed (i.e. one blastomere from the two-cell stage).

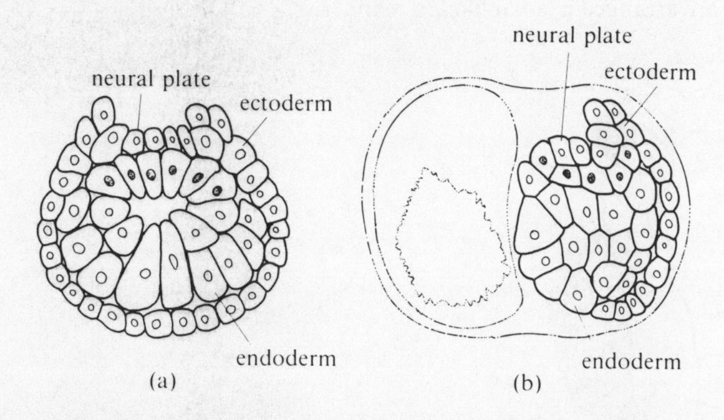

FIGURE 1 (a) Cross-section through a normal *Styela* embryo at the neural plate stage. (b) Cross-section through a *Styela* embryo derived from the two-cell stage in which one blastomere was killed.

□ Has any determination occurred in *Styela* by the two-cell stage?

■ Yes, it appears so, as a single live blastomere yields only a half-embryo later on. That is, an embryo that, although it appears intact when viewed externally, in fact, lacks all the internal organs on one side (the side corresponding to the dead blastomere).

So the single live blastomere was incapable of generating an intact embryo later on. This suggests that the bilateral (left–right) symmetry of *Styela* is fixed by the two-cell stage. As some aspects of the fates of the cells are determined by the two-cell stage, we are almost forced to conclude that the fertilized egg itself contained some local differences, so that when it divided to give two cells, each inherited a different set of components. These differences presumably ensured the different fates of the two cells. In fact, examination of a fertilized *Styela* egg reveals differences in local contents so that the egg is differently pigmented in different

* The Open University (1981) S202 SLO *A Survey of Living Organisms*, The Open University Press. This text forms part of the supplementary material of the Course.

regions. The fates of these different pigments can be followed during the development of the embryo and, as Conklin found, they can always be located in particular organs at a relatively advanced embryonic stage (Figure 2).

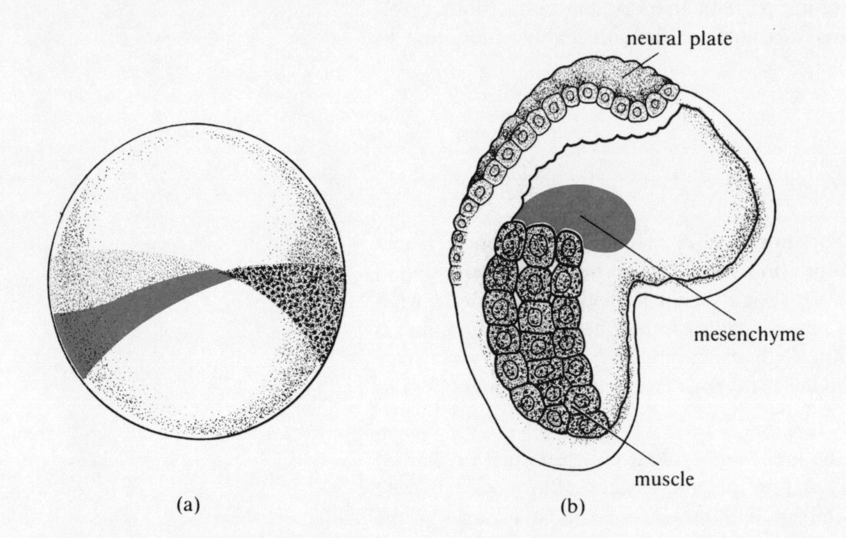

FIGURE 2 (a) Pigmentation in a fertilized *Styela* egg. (b) Pigmentation of tissues in a later *Styela* embryo.

This suggests that particular regions of the egg always give rise to particular organs.

□ Does this necessarily mean that these pigments are themselves responsible for the formation of particular organs?

■ No. But the experimental observations produced the idea that there was localization of 'organ-forming' substances in different regions of the egg.

Though not conclusive, the fate of the pigments, and Conklin's experiments described in Figure 1, suggest that the egg of *Styela* is a '*mosaic*'; that is, each part of the egg is responsible for a particular part of the adult. This view is further supported by some experiments, shown in Figure 3, in which a fertilized *Styela* egg is centrifuged during the first cell division and then allowed to develop.

mosaic development

By comparing Figure 3 with Figure 2 you should be able to see that this results in an embryo in which the tissues and organs are arranged in a haphazard manner.

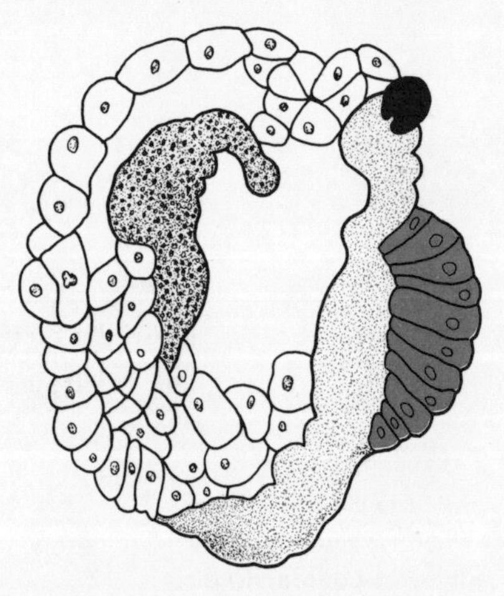

FIGURE 3 Cross-section through an embryo derived from a fertilized *Styela* egg which was centrifuged during the first cell division.

localization

This would suggest that the *localization* of certain factors within the egg is responsible for the subsequent production of differentiated cells in the correct spatial positions. Upsetting the location of these factors upsets the subsequent development of the cells.

All these experiments point to there being a localization of certain factors of developmental significance in the fertilized *Styela* egg, so that during the cell divisions these factors are differentially partitioned among the various cells; this then leads to the differential development of those cells.

This idea echoes the 'preformationist' view of development that was popular in the eighteenth century, wherein an egg was supposed to contain a preformed adult, that is a mosaic pattern whereby each minute part of the egg contained a preformed version of a corresponding part of the adult. The extreme version of this view was advanced by Bonnet in 1745. He supposed that each egg contained a pattern for the formation of any embryo, which in turn contained eggs containing smaller embryo patterns, and so on. Following this line of argument, Bonnet calculated that Eve must have contained 27 000 000 embryos within embryos in her ovaries!

Though we may now scoff at the naivety of such ideas, the experimental results from using *Styela* still have to be explained. Throughout the *Development* Block we have taken the view that development proceeds by *epigenesis* (i.e. a progressive series of changes whereby complex systems are constructed from less complex ones) rather than seeing development as resulting from some preformed embryo in the egg. The onus is therefore on us to fit the results obtained with *Styela* into an epigenetic framework. Perhaps the first question is: how typical is *Styela*?

epigenesis

2.2 Regulative development

As in Conklin's experiments on *Styela*, in many organisms it can be shown that determination is an early event in embryogenesis, which seems to reflect some mosaic structure in the egg.

□ Can you give an example to show that in some organisms the blastomeres might *not* be determined by the two-cell stage?

■ In humans, for example, monozygotic twins can arise from a single fertilized egg. This might involve separation of the two blastomeres of the two-cell embryo, or separation of cells at an even later stage. Yet, unlike *Styela*, such separated cells can give rise to normal individuals.

Many such examples are known. In the sea-urchin any one of the blastomeres of the two-cell or four-cell stage, if isolated, can give rise to a complete pluteus (a larval stage in sea-urchin development). This means that at the two-cell stage an isolated, left-hand blastomere, for example, gives rise to an embryo with both left-hand and right-hand organs. Other similar cases are known, where once again a perfect, intact, embryo is formed from a single isolated blastomere from the two-cell stage, though sometimes the embryo thus obtained is smaller than normal. (Indeed, in the sea-urchin, if a complete blastula is cut in half longitudinally, then each half can round-up, and then gastrulate and go on to form a qualitatively complete pluteus.)

So it appears that in some organisms, such as *Styela*, the blastomeres are determined early on, thus fixing bilateral symmetry, but in other organisms, such as the sea-urchin, they are not. The latter organisms are said to be *regulative*, as it is considered that their development is not predetermined by any mosaic in the egg, but that each blastomere is capable of some degree of regulation to fit its new surroundings. For example, a left-hand blastomere of a two-cell stage if isolated from a regulative organism can adjust its normal role to take over the roles of both left-hand and right-hand blastomeres, thus producing a normal embryo. Similarly, if two embryos of a regulative organism are fused before total determination has set in, they can go on to produce a normal adult rather than a 'double-organned monster'. For example, two mouse eight-cell embryos can be fused to yield finally an adult mouse. Presumably in this case each cell in each eight-cell embryo had sufficient ability to regulate itself in response to the new situation to produce an essentially normal 16-cell embryo rather than some 'twinned monster'.

regulative development

It is now worth considering the apparently logical consequence of the phenomenon of regulation. We argued above (Section 2.1) that early determination in *Styela* suggested that the egg was a mosaic, that is, within the egg there were localized factors that profoundly and differentially influenced the developmental fate of the individual blastomeres that inherited them. *Therefore one might conclude that in regulative organisms the eggs contained no such localized factors.* So, are we to assume that two basic types of development exist? One, the mosaic type, in which the egg contains localized factors which inevitably determine the fates of the cells to which they are apportioned; the other, the regulative type, in which there is no localization in the egg and the developmental fate of the cells is gradually determined via interactions between cells.

The answer to this question comes from some interesting experiments on sea-urchin eggs and a closer examination of mosaic organisms such as ascidians (e.g. *Styela*).

2.3 Localization in regulative eggs

The normal development of a sea-urchin egg up to the stage of the pluteus is shown in Figure 4. You will recall this from Units 12 and 13.

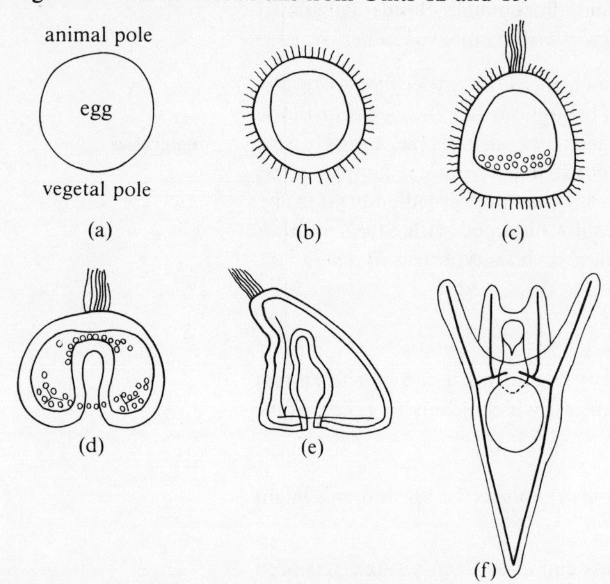

FIGURE 4 Schematic representation of sea-urchin development from egg to pluteus: (a) the egg; (b) and (c) early/late blastulae; (d) the gastrula; (e) the prism; (f) the pluteus.

As we have seen, a single isolated blastomere from the two-cell or four-cell stage, or even a half-blastula, can produce an intact pluteus. However, in the 1930s Hörstadius performed some elegant and interesting experiments, the results of which at first sight seem inconsistent with these findings.

What Hörstadius did was to take sea-urchin eggs and cut them in half in a *horizontal plane*, that is, between the animal and vegetal poles. In this way he obtained animal and vegetal half-eggs. The halves could each be fertilized and he thus obtained haploid or diploid half-eggs, depending on whether they contained the sperm nucleus alone or both the sperm and egg nuclei. But the subsequent development of any one of the fertilized half-eggs was the same irrespective of whether it was haploid or diploid. What did differ, as can be seen from Figure 5, was the fate of the animal and vegetal halves.

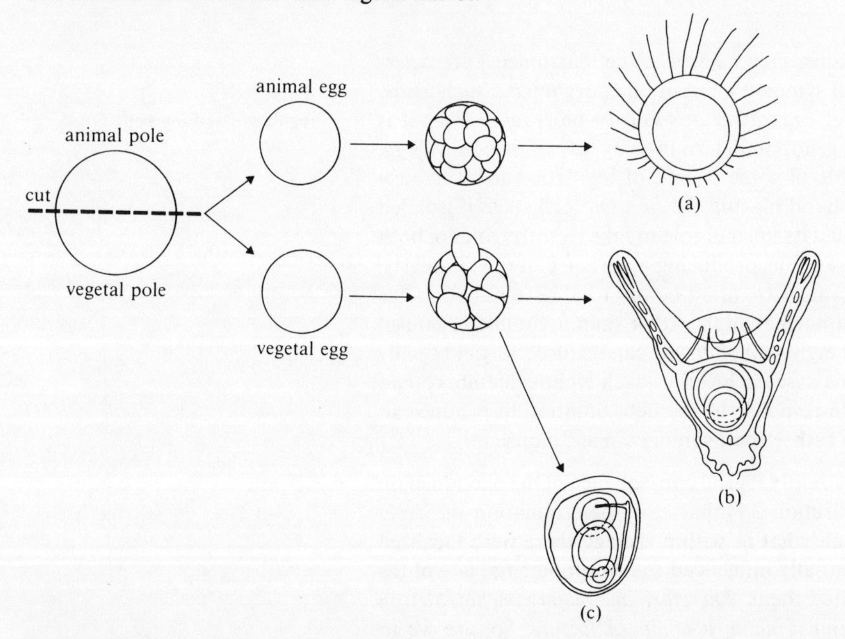

FIGURE 5 Hörstadius's experiment. Each half of the egg (animal or vegetal) rounds-up, can be fertilized, and develops as shown. The animal half shows little change in overall shape (a). The vegetal half can yield an essentially normal pluteus (b) or an aberrant one (c).

The different potentials for development of animal and vegetal half-eggs cannot be due to differences in their chromosome content as the results are not affected by the number of nuclei in the half-eggs.

□ Do these experiments support the idea that localization exists in the regulative sea-urchin egg?

■ Yes. They obviously reflect some local differences between the animal and vegetal halves of the egg.

This does not necessarily imply that sea-urchin eggs contain 'organ-forming' substances in the way that *Styela* eggs apparently do. In addition, centrifugation of sea-urchin eggs does not lead to the production of abnormal larvae, as it did in *Styela*. Nevertheless, Hörstadius's experiments do imply some local differences somewhere in the sea-urchin egg, though these may be of a 'broader' nature than in *Styela*, affecting the overall animal as opposed to vegetal characteristics.

So in a regulative organism also, localization does exist in the egg. The same appears to be true for other regulative organisms such as amphibia. Thus, the mosaic or regulative nature of an organism does not argue for or against localization in the egg. It is probable that some degree of localization of factors of developmental significance occurs in all eggs. In fact, the difference between mosaic and regulative eggs is much less distinct than it once appeared to be: no organism is wholly regulative or wholly mosaic. For example, even in *Styela*, which was chosen as a classical mosaic organism, some regulation can occur at early developmental stages. If each blastomere from a two-cell embryo is separated from the other (rather than just killing one), then each isolated blastomere can develop to yield *somewhat more than a half-embryo*. So some regulation must be possible.

Compare this with Conklin's experiment (Figure 1) from which it seems that early determination is an important event in the development of *Styela*. As the experiment in which blastomeres *are actually separated* shows, the attached dead blastomere must also play some part in preventing the live one from regulating. How it does this we do not know. One possibility is that the live blastomere is unable to reconstitute an intact exterior surface while the dead one is attached (perhaps because of mechanical interference). So, as well as 'organ-forming' substances in the interior of the egg, the surface structure of the egg or early embryo may also be important in organizing subsequent development. Alternatively, substances present in the attached dead blastomere may inhibit the partial regulation that could otherwise occur in the live blastomere.

If we now accept that mosaic and regulative organisms both exhibit structural organization in the egg, and that this affects subsequent development of individual blastomeres, this leaves us with several new questions:

1 How is it that an isolated blastomere from a sea-urchin embryo can develop normally, yet a half-egg, as produced by Hörstadius, cannot?

2 What is the nature of the factors that are localized in the egg?

3 How do these factors affect development?

4 How does localization of the factors arise?

In the following Sections we shall attempt to answer these questions, beginning with the curious paradox presented in question 1.

Objectives and SAQs for Section 2

Now that you have completed this Section, you should be able to:

★ evaluate evidence or statements about whether the development of an organism is of the regulative or mosaic type.

★ decide on the validity of data bearing on the notion of localization in particular eggs.

To test your understanding of this Section, try the following SAQs.

SAQ 1 (*Objective 1*) Match each term in List A with its correct definition in List B.

LIST A

(a) localization
(b) preformationism
(c) mosaic development
(d) regulative development
(e) determination
(f) epigenetic development

LIST B

(i) Development in which individual cells depend on interactions with other cells in order to differentiate normally and in which cells can adjust to compensate for changes from the norm.

(ii) The phenomenon in which cells at particular stages in their differentiation reach an absolute point of no return so far as their developmental fates are concerned, even if the environment of the cells is experimentally altered.

(iii) The phenomenon in which factors influencing cell differentiation are differentially distributed in the egg.

(iv) Development that depends on the progressive generation of complex structures from less complex ones.

(v) The idea that the complete pattern for each part of the adult is contained in the egg.

(vi) Development in which individual blastomeres are fixed in their developmental fates very early on, irrespective of whether they are isolated from the whole embryo or not.

SAQ 2 (*Objectives 1 and 2*) Is it true that cell–cell interactions are likely to be relatively more important in the early development of regulative organisms than in mosaic ones?

3 The plane of cleavage

The experiments of Hörstadius imply that there is a qualitative difference between the animal and vegetal halves of the sea-urchin egg. If these differences, which appear to be relevant to subsequent development (see Figure 5), are maintained during cleavage then we would expect that blastomeres arising from different parts of the original egg substance would show different developmental fates. Yet, paradoxically, blastomeres at the two-cell or four-cell stage all appear equivalent, any one blastomere being capable of yielding a normal embryo.

The simplest way to resolve this paradox is to assume that the first two rounds of cell division (i.e. the first two cleavages) of the sea-urchin egg must divide the egg substance in such a way that each resulting blastomere contains a similar mix of any localized factors of developmental importance. On the other hand, the 'artificial cleavage' performed by Hörstadius must result in an unequal partitioning of such factors between the two half-eggs. Hörstadius's surgical 'cleavage' cuts the egg into the animal and vegetal halves; the first two rounds of normal cleavage in development are known to divide the egg longitudinally and symmetrically. Thus, if we *assume* that sea-urchin eggs contain stratified regions of developmental factors at right angles to the animal–vegetal axis, we obtain the situation in Figure 6.

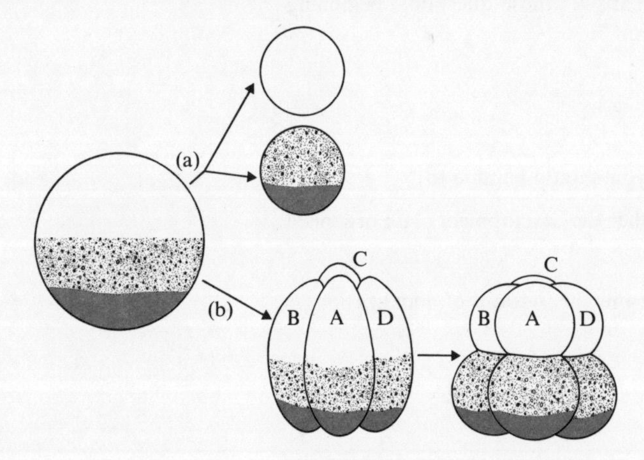

FIGURE 6 Schematic representation of postulated strata in sea-urchin eggs and their fates during cleavage. (a) Partitioning of strata as a result of surgical 'cleavage'. (b) Partitioning of strata as a result of normal cleavage (two-cell stage omitted).

Figure 6b shows that all four blastomeres (labelled A–D) in the four-cell stage contain an equal mix of the strata (shown as different colours and shades) and hence are equivalent. By the third cleavage, qualitative differences would have arisen and this ties in with observations that some degree of determination has occurred by the eight-cell stage in the sea-urchin. However, the surgical 'cleavage' (Figure 6a) would result in immediate differences between the two halves.

□ Assume that the above conclusions are valid, and bear in mind Hörstadius's experiments (Figure 5). What would you expect to happen to two fertilized half-eggs produced by surgically cutting the egg along the vertical axis and then fertilizing the halves?

■ Each half should develop to give a complete pluteus. This does, in fact, happen, which serves to show that the cutting of the egg itself does not cause the result shown in Figure 5. This reiterates that it is the direction of cutting and the concomitant partitioning of localized structures in the egg that are crucial.

In other organisms the early cleavages are not symmetrical and a stratified egg would give a different result (Figure 7).

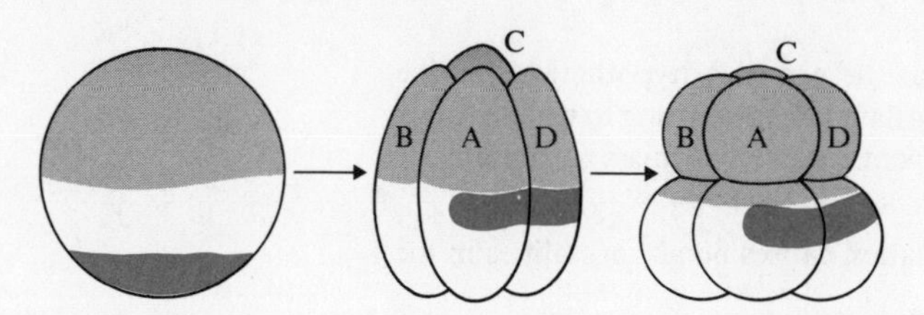

FIGURE 7 Schematic representation of postulated strata and their fates in an egg undergoing asymmetric division (two-cell stage omitted).

By comparing Figure 7 with Figure 6b you can see that in Figure 7, at the four-cell stage, the four blastomeres (A–D) are not all equivalent, as only A and D contain the red band (and only B and C contain other strata, which are not shown). Hence in this instance by the four-cell (A–D) stage (and even at the two-cell stage) the blastomeres could already differ in their contents of factors originally localized in the egg. A situation of this sort presumably exists in *Styela*.

Summary of Section 3

1 The localization of factors of developmental significance probably occurs in all eggs.

2 The planes of the early cleavages determine whether the factors are distributed evenly among the blastomeres or not.

3 The even distribution of factors ensures that the blastomeres can, *at least in principle*, remain equivalent.

So, sooner or later, in all organisms, cleavage results in an uneven partitioning of developmental factors among the blastomeres. If this occurs early on, the development appears to be mosaic (e.g. *Styela*); if it occurs later (e.g. between the four-cell and eight-cell stages in the sea-urchin), development can in principle be regulative. That is, a blastomere isolated at any time up to the stage when cleavage becomes asymmetric with respect to the developmental factors, can in principle yield the whole organism. So, whether one classifies an organism as mosaic or regulative has no real bearing on the existence or otherwise of localized factors in the egg: historically, however, it was easier to suppose and support the idea of localization in mosaic organisms.

Mosaic and regulative organisms differ in the *timing of cell determination*. They also differ in the fixing of axes of symmetry which presumably depends on how precisely cleavage partitions factors localized in the egg.

timing of determination

4 The nature of the localized factors

Virtually nothing is known about the chemical nature of the different factors postulated to exist in the various regions of the egg. We can conclude from the centrifugation experiments on *Styela* (Figure 3) that in this organism at least, whatever the factors are, their position in the embryo can be disturbed by centrifugation. However, this is not true in all organisms and in some, such as amphibians, it appears that the factors are located in the stiff outer layer of the egg—the cortex. (Note that the surface of the early embryo may also be important in *Styela*; Section 2.3.) The localization of factors, whatever they are, need not be qualitative. That is, it need not depend on different factors being located in different positions in the egg. It may instead be wholly or partly quantitative and depend on different levels of one or more factors throughout the egg. That is, there may be *gradients* of one or more factors in an egg that help to produce the subsequent changes. Evidence for gradients of this sort comes from studies on sea-urchin eggs.

gradients in eggs

Objective and SAQ for Sections 3 and 4

Now that you have completed these Sections, you should be able to:

★ decide whether data support or refute the notion of localization in particular eggs.

To test your understanding of these Sections, try the following SAQ.

SAQ 3 (*Objectives 1, 2 and 3*) Imagine that the egg of a hypothetical marine organism is under study. Classify the following data (a-c) according to whether they support or refute the notion that development in this organism is mosaic, or whether the data do not discriminate.

(a) Centrifugation of the egg just after fertilization causes no abnormalities in the subsequent development.

(b) Removal of one blastomere from the two-cell embryo results in highly abnormal development by the remaining blastomere, the later embryo lacking several structures.

(c) Addition of certain nitrated phenols to an egg soon after fertilization causes abnormal development.

5 Factors and cell differentiation

So far, we have seen that in the egg, either in the cytoplasm or the cortex, there is an uneven distribution of factors that influence the development of cells subsequently formed. Because of the uneven distribution of these factors among the daughter-cells (blastomeres), resulting from the early cleavage of the egg, these cells develop in different ways.

The question now is: what do these factors do to cause this differential development?

In Units 12 and 13 we considered that cellular differentiation involved the differential transcription[3] of genes. In principle, this could involve a difference in genetic information between the cells or a switching-on/switching-off of different genes from among the same set. The experiments of Steward and of Gurdon (Units 12 and 13) led us to favour the latter alternative.

Applying this to our current problem, we would suppose that all the blastomeres contain the same complement of genes (i.e. they are in this respect totipotent) and that the differences in the factors present in the different blastomeres cause, directly or indirectly, the switching-on/switching-off of different genes. From the work of Gurdon and Harris (Units 12 and 13), you are already familiar with the concept that genes in the nuclei will behave differently, depending on the cytoplasm in which those nuclei are located. Thus we can now equate what we earlier called cytoplasmic signals with what, in the context of the egg, we call developmental factors. Other factors may be located in various portions of the original egg surface, so that cells inherit these differentially, too. Some such signals (factors) may also of course regulate mRNA[4] translation[5] (Units 12 and 13); others perhaps regulate gene transcription. *We may therefore conclude, albeit tentatively, that the start of epigenesis depends on the differential development of different blastomeres, and that this, in turn, depends on the uneven distribution of certain developmental factors among the blastomeres.*

6 The origins of localization

Accept for the moment that we have produced a plausible explanation of how the early changes in the developing embryo start—the question we posed at the beginning of this Unit. In other words, accept that they start because of the influence of factors present in the early embryo or egg, which are localized in certain regions.

How does this localization arise?

The short answer is that no one knows for certain; what is known is that localization varies from species to species. We can occasionally tell something about *when* localization arises; for example, a *Styela* egg centrifuged just after fertilization still gives rise to a normal embryo. Presumably in this organism localization occurs a little later on (Figure 3). One relatively simple manifestation of localization about which certain things are known is the phenomenon of *polarity in the egg*.

polarity (of egg/embryo)

6.1 Polarity in the egg

One of the earliest manifestations of development in both animals and plants is polarity: the shoot and root poles in plants, or the anterior and posterior body poles in animals. Even the first cleavage of the initial zygote cell is usually oriented with respect to some axis. This orientation must presumably reflect some localization within the zygote cell, and so eggs showing polarity can be regarded as useful model systems for asking how localization in general arises.

In some organisms, such as the seaweed *Fucus*, the polarity in the egg is very labile (unstable). About 17 hours after fertilization and still before the first cleavage, the *Fucus* egg starts to grow a long projection, called a rhizoid, on one side of the egg. This indicates a polar axis. Generally, the rhizoid grows out at or near the point where the sperm earlier entered the egg. However, the point at which the rhizoid actually grows out can be profoundly altered by subjecting the egg, before about 15 hours after fertilization, to changes in pH, or illumination on one side, or treatment with a plant growth hormone, or even by placing it in an electric field. Thus in this organism, polarity is not stabilized, or perhaps not established till about 15 hours after fertilization.

In some other organisms polarity is much more stable and is determined before fertilization. For example, in *Ascaris megalocephala* (a nematode) the polarity of the egg seems to be determined by its position as it develops in the ovary, so that the end of the egg that eventually becomes the animal pole was the one that was against the oviduct wall. In sea-urchins it appears that the vegetal pole corresponds to the end of the egg that was attached to the ovary wall.

So, it appears that polarity in an egg, which presumably reflects some localization within that egg, in some instances depends on the developmental history of the egg itself within the mother.

We argued at the beginning of this Unit that to explain how 'change starts', and hence to explain how development is set on its course, it is necessary to investigate the structure of the egg. Now, curiously, in order to explain the structure of the egg, we may need to consider its origins too. Which in one sense leaves us with ...

SAQ for Sections 5 and 6

To test your understanding of these Sections, try the following SAQ.

SAQ 4 (*Objectives 1, 2 and 3*) Which of the following statements are *true* and which are *false*?

(a) The plane of the first cleavage in mosaic organisms is always different from that in regulative organisms.

(b) The polarity of eggs is unalterable.

(c) In mosaic organisms early cleavages lead to an uneven distribution of factors important to development among the resulting blastomeres.

(d) Factors important to cellular differentiation are always located in the cytoplasmic fluid of the eggs.

7 Development: a summary

In this Unit, as in our treatment of development as a whole, we have taken particular views of development. These views are by no means the only ones, and could reasonably be argued against. However, given the confines of five Units, it has been necessary to adopt particular views and hence to select particular examples bearing on these views. This has inevitably meant ignoring different modes of development in other organisms. Whether or not these differences, which exist in the developmental strategies of different organisms, merely reflect differences in detail and degree or more fundamental differences is a matter of debate.

Our discussion of development has also ranged over a large number of topics and sometimes the links between them may have seemed somewhat tenuous. This is probably because of the rather uncertain nature of the subject of developmental biology today, and the difficulties of writing such a compact treatment. So we shall now very briefly summarize some major events in the development of a typical adult animal from a fertilized egg. For a change, we shall consider this in terms of the sequence of events, but again we will emphasize the component processes hitherto discussed (Units 11–15). (You should bear in mind the impossible nature of the task of ever describing a 'typical' animal and the speculative elements inherent in any such summary.)

7.1 From egg to adult

The fertilized egg contains a nucleus within which are the genes. It also contains, localized within it, a differential distribution of factors that can affect the production of specific proteins by regulating gene transcription and possibly mRNA translation. The egg divides (cleaves) to give two blastomeres. Cleavage continues; each blastomere contains a complete copy of the genes present in the fertilized egg (Units 12 and 13). Sooner or later cleavage results in an uneven distribution to the blastomeres of the developmentally important factors localized in the egg (Unit 15). These differences result (sooner or later) in different proteins (Units 12 and 13) being produced in different blastomeres because of a combination of differential gene transcription and controlled translation of mRNA (Units 12 and 13). The blastomeres thus begin to differ qualitatively from each other—cell differentiation begins (Units 12 and 13). Now each blastomere finds itself in a unique local environment as it differs from its neighbouring cells. These local differences may lead to further differentiation of existing and subsequently produced cells. The molecular basis of these changes involves signals of a chemical or physical nature between cells (Unit 14) and subcellular responses to these signals (Units 12–14). This further cell differentiation leads in turn to new local environments, hence new signals, and so on. The signals may be of an all-or-none nature or of a quantitatively graded nature (Unit 14). The grosser manifestations of the responses to these signals are changes in cell shape and size and even in the movement of cells (Unit 14). These changes lead to the phenomena known as embryonic induction, pattern formation and morphogenesis (Unit 14). In some instances genes may enable cells to act together with other cells to form patterns, as in the case of polyclones (Unit 14). The changes are of course coordinated (Unit 11) and under temporal control. This temporal control may be both at the subcellular level, where sequences of changes may occur within one cell (Units 12 and 13), and at the intercellular level, where the phenomena of competence and embryonic induction (Unit 14) also impose a time sequence. All these processes act together in accordance with the original genetic programme in the egg and the interaction of this programme with the environment, to produce finally an adult organism (Unit 11). Changes in either the programme or environment can lead to altered development.

The above summary is very sketchy, and this reflects the difficulty in producing a linear argument for a system in which many processes are operating simultaneously and are interdependent. There is also the difficulty, foreseen in Unit 11, that it may be meaningless to consider 'development' as a single phenomenon; it may have a precise meaning only when applied to a particular organism. However, in this *Development* Block we have subscribed to the view that it is meaningful to consider the component processes of development without having to consider the complete development of any one organism. It is probable that these component processes underlie the development of all organisms (Unit 11), but the degree to which each is employed in the development of an organism may vary from species to species. To take an example: in regulative organisms cell–cell interactions are very important in the process of early determination because a blastomere in an intact embryo will develop to produce its normal organs, but the same blastomere if isolated from its neighbouring cells will behave differently and regulate. In mosaic organisms this is not so and here the distribution of localized factors in the egg is relatively more important than cell–cell interactions. However, no organism is wholly mosaic or wholly regulative: so-called regulative organisms do depend on some localization in the egg, and mosaic organisms may depend on cell–cell interactions for later developmental changes (e.g. the formation of the nervous system in ascidians depends on embryonic induction).

7.2 The questions that remain

The purpose of this *Development* Block has been to introduce you to some of the concepts underlying developmental biology and thus enable you to pose more precise questions. The above summary belies the problems of developmental biology that remain to be solved. Some of these will be obvious from the individual Units, but let us end our discussion by highlighting a few of the remaining questions:

1 How are genes switched on or off during cell differentiation? Are the mechanisms of the Jacob–Monod type, or not? (Units 12 and 13)

2 How is the structure of the eukaryote chromosome involved in differential gene transcription? Is folding and unfolding of chromosomes involved? (TV programme *Differential gene expression*)

3 What is the nature of the signals between nucleus and cytoplasm? Do these signals help maintain the differentiated state of the cell, and if so how? (Units 12 and 13)

4 What is the nature of the signals between cells? Are the signals involved only in local interactions, as in embryonic induction, or do more global signals also exist, as suggested by theories of positional information?

5 How do cells interpret signals and, in particular, how can cells interpret the level of a signal in a gradient? (Unit 14) Does interpretation always involve gene activity leading to the synthesis of new proteins?

6 What controls cell movement? (Unit 14) What is the relationship between external factors and internal cell responses resulting in cell movement or its inhibition?

7 Does membrane fluidity play a part? How is the cell surface involved in cell movement and adhesiveness? (Units 5 and 14).

8 Does differential cell adhesion assist in morphogenesis and, if so, how? (Unit 14) Is sorting out due to differential cell adhesion? Is differential cell adhesion a means of morphogenesis, or just a consequence of it?

9 How does localization in eggs arise? (Unit 15) What are the relative roles of environment and maternal genotype in producing localization?

10 How are the planes of cleavage in the egg determined? (Unit 15) What are the roles of genotype and environment in determining these planes?

11 What is the basis of the organizing nature of the egg or embryo surface? (Unit 15) How does this relate, if at all, to the fluid nature of membranes?

7.3 Conclusions

The classic experiments of embryologists from about 1890 to 1940 demonstrated the crucial importance of cell–cell interactions during development. The molecular biologists of the 1950s and 1960s helped unravel the problems of how genes might influence development and how their activity might be controlled. It is likely that the breakthroughs in the next decade or so will be at the interface of those two areas: how cells receive and interpret signals from the external environment and other cells, and how they act on these signals to produce the carefully regulated subcellular changes. The actual *physical interface*, the cell surface, is likely to be crucial to those processes (see questions above); the cell surface therefore promises to be the most studied biological structure of the 1980s.

Objectives for Unit 15

Now that you have completed this Unit you should be able to:

1 Define and use, or recognize definitions and applications of all the terms, concepts and principles asterisked in Table A.

2 Evaluate evidence or statements about whether the development of an organism is of the regulative or mosaic type. (*SAQs 2, 3 and 4*)

3 Decide whether data support or refute the notion of localization in particular eggs. (*SAQs 3 and 4*)

SAQ answers and comments

SAQ 1 (a) (iii); (b) (v); (c) (vi); (d) (i); (e) (ii); (f) (iv)

SAQ 2 Yes, virtually by definition, because in regulative development, each early blastomere can, if isolated, yield a whole embryo, while if it is left intact in the embryo it contributes only some of the tissues and organs. This means that the behaviour of such a blastomere depends on its interactions with its neighbouring blastomeres (e.g. by embryonic induction). In mosaic development the behaviour of each blastomere seems less influenced by whether it is on its own or in contact with the other blastomeres.

SAQ 3 (a) This does not discriminate. For example, development could be mosaic but the 'factors' could be in the cortex (Section 4).

(b) Supports the notion; as in *Styela* (Section 2.1).

(c) This does not discriminate; introduced chemicals *could* potentially upset any type of development.

Note that (a) and (c) do not discriminate and only (b) *really* supports the idea of early determination. The difficulty of deciding between *truly* mosaic and regulative development is a matter of definition; as seen in Section 3 the distinction between the two modes of development is actually rather arbitrary.

SAQ 4 (a) False. The plane may be in the same geometric direction. However in mosaic eggs this plane presumably partitions the factors localized in the egg in an uneven manner. In regulative eggs the planes of early cleavage are more symmetrical in this respect.

(b) False (e.g. *Fucus*).

(c) True (see answer to (a)).

(d) False. They may be in the cortex (Section 4).

References to the Foundation Course

S101

1 Unit 18, *Natural Selection*, Section 1.

2 Unit 25, *DNA, Chromosomes and Growth: Molecular Aspects of Genetics*, Introduction and Section 8.

3 Unit 25, Section 4.2.

4 Unit 25, Section 4.

5 Unit 25, Section 4.3.

S100

1 Unit 17, Section 11.

2 Unit 17, *The Genetic Code: Growth and Replication*, Section 13.

3 Unit 17, Section 6.

4 Unit 17, Sections 6 and 7.

5 Unit 17, Section 7.

List of Units

The Diversity of Organisms

Unit 1	Marine Organisms
Unit 2	From Sea to Land: Plants and Arthropods
Unit 3	From Sea to Land: Vertebrates

Cell Biology

Unit 4	Cell Structure
Unit 5	Macromolecules and Membranes
Unit 6	Enzymes: Specificity and Catalytic Power
Unit 7	Enzymes: Regulation and Control
Unit 8	Metabolism and its Control
Units 9 and 10	Membrane Transport and Energy Exchange

Development

Unit 11	Development: The Component Processes
Units 12 and 13	Cellular Differentiation
Unit 14	Pattern Specification and Morphogenesis
Unit 15	Chicken or Egg?

Animal Physiology

Unit 16	Communication: Nerves and Hormones
Unit 17	Blood Sugar Regulation
Unit 18	Control Mechanisms in Reproduction
Unit 19	Circulatory Systems
Unit 20	Respiratory Mechanisms
Unit 21	Respiratory Gases and Their Transport
Units 22 and 23	Osmoregulation and Excretion
Units 24 and 25	Nutrition, Feeding Mechanisms and Digestion

Plant Physiology

Unit 26	Plant Water Relations
Unit 27	Ion Movement and Phloem Transport
Unit 28	Plants and Energy
Unit 29	Plant Cells: Growth and Differentiation
Units 30 and 31	Morphogenesis in Flowering Plants